Development of Economic Thought

The Authors

Prof. Victor Adeyeye Abimbola graduated from University of Ife in B.Sc (honours) in agriculture in 1978 and M.Sc. in Agricultural Economics in 1980 from University of Ibadan, Nigeria. He earned Ph.D. degree in Agricultural Economics from University of Ibadan, Nigeria in year 1986. Professor V.A. Adeyeye has published more than 150 research papers in field of Agricultural Economics and Economics. He shared the responsibility in organizing several National Seminars on Social Science and Applied Economics Research. At present, he is working as a Research Professor and Director in the Department of Economic Policy Unit, Nigeria Institute of Social & Economics Research (NISER), Ibadan, Oyo State, Nigeria.

Dr. Olayiwola Olubodun Olaniyi was hail from Ilora community in Afijio Local Government of Oyo State, Nigeria. He completed his Ph.D in the year 2015 under the sponsorship of African Union Commission. Dr. Olayiwola Olubodun Olaniyi papers have been referred in various text and reference books published both national and international publishers.

Development of Economic Thought

Victor Adeyeye Abimbola

Olayiwola Olubodun Olaniyi

2018

Daya Publishing House®

A Division of

Astral International Pvt. Ltd.

New Delhi – 110 002

Cataloging in Publication Data--DK
Courtesy: D.K. Agencies (P) Ltd. <docinfo@dkagencies.com>

Adeyeye, V. A., author.
Development of Economic Thought / Victor Adeyeye Abimbola, Olayiwola Olubodun Olaniyi.
pages cm
Includes bibliographical references and index.
ISBN 9789387057579 (International Edition)

1. Economics--History. I. Olaniyi, Olayiwola Olubodun, author. II. Title.

HB75.A34 2017 DDC 330.09 23

Published by : **Daya Publishing House®**
A Division of
Astral International Pvt. Ltd.
– ISO 9001:2015 Certified Company –
4736/23, Ansari Road, Darya Ganj
New Delhi-110 002
Ph. 011-43549197, 23278134
E-mail: info@astralint.com
Website: www.astralint.com

Dedicated to

I dedicate this book to God Almighty and entire family, friends and other scholars for their contribution towards the success of the book.

Foreword

The goal of the managers in the economy is always to close the gap between the potential and actual GDP, that is, remove the bottlenecks and create a situation for full employment and economy stability. It is an established fact that backward societies have not succeeded in making full use of their potentials for development not only because of the underdevelopment of their natural resources, including human resources and production technologies, but also due to the underdevelopment of the existing socioeconomic organization and institutions, including the network of markets and administrative machinery of the government in such societies. The general presumptions therefore, is that the elimination of these structural deficiencies in the economy of a backward society would enable such society, be it a nation, state or local government area, to make better use, not only of its domestics resources, but also its external economic opportunities in the form of international trade, and technological and organization innovations. It is in the pursuance of the foregoing that developing countries have always engaged in economic growth and development planning. The conventional plan attempts to promote economic growth and development in three main ways, namely:

(a) Increasing the overall rate of investment in the economy

(b) Engaging in some special type of investment designed to ease the bottlenecks of production in the important sector of the economy

(c) Trying to improve the coordination between different sectors and sub-sector of economy.

Lastly, the field of Agriculture economics would provide guidelines for making adjustment in the use of scares resources in a rational direction and prevent expending of resources in wrong direction and provide appropriate plans for government to execute their developmental policies and programmes.

Victor Adeyeye Abimbola

Preface

The book entitled "**Development of Economic Thought**" has been written for the entire post graduate, under graduate and PhD students of social science. The field of economics is wide and characterized by tortuous terrain. Although there are several texts and publications in economics, books which focus on economics thought but they are not precise enough. In order to close the knowledge gaps, the structure of the book is design to give more details about the economic thought ideas in a simplest ways. The book has been compiled bearing in mind a variety of scholars needs that focused on developmental issues. I must add here that the bulk of materials in this book had come in the main from a compilation of field surveys and Focus Group discussions, personal Interviews, questionnaire and literature reviews on the subject matters.

Victor Adeyeye Abimbola

Olayiwola Olubodun Olaniyi

Contents

1
History of Economic Thought

1.1 History of Economic Thought

The subject, the History of Economic Thought, may be defined as a critical account of the development of economic ideas, searching into their origins, interrelations, and, in some cases, their results. The history of economic thought concerns thinkers and theories in the field of political economy and economics from the ancient world right up to the present day. Economics was not considered a separate discipline until the nineteenth century. For example, Aristotle, the ancient Greek philosopher, in his works on politics and ethics has thought of 'art of wealth acquisition'. He also considered the question whether property is best left in private or public hands. In medieval times, scholars like Thomas Aquinas argued that it was a moral obligation of businesses to sell goods at a just price. Economic thought evolved through feudalism in the middle Ages to mercantilist theory during the Renaissance (when people were concerned to orient trade policy to further the national interest). The modern political economy of Adam Smith appeared during the industrial revolution, when technological advancement, global exploration, and material opulence that had previously been unimaginable were becoming a reality. All these and further developments are subject matter of history of economic thought. Changes in economic thought have always accompanied changes in the economy, just as changes in economic thought can propel change in economic policy. Economic thought has at times focused on the aspects of human nature such as greed and selfishness that generally work against the good of all; at other times, economic behaviour has been seen as self-regulating and working toward a common purpose. As contemporary economic thought deals with the issues of globalization and the emergence of a global economy, economists have turned to the multitude of other disciplines which, like economics, developed independently. Building on their discoveries, and united with them in pursuit of the common goal of benefiting human society, economic thought may be on the road to achieving a new level of understanding. There are several ways to present the history of economic thought:

- ✰ To analyze the changing nature of economic theory in conjunction with the social and economic development of society
- ✰ To emphasize economic thinking as part of the main currents of philosophical and political ideas
- ✰ To emphasize the internal dynamics of the science where new insights and results emerge as a consequence of economists' awareness of the shortcomings of the present state of the subject.

1.2 Economic History

Economic history is different from history of economic thought. Economic history is the study of the economic aspects of societies in the past; the history of the economic use of resources land, labour and capital; or the examination of the past performance of economies. It is concerned with how people lived most of their lives, how many were born and died, how they earned and spent, worked and played. Such variants, however, reveal little more than the definition which once said simply that it was the sort of history which required a knowledge of economics; though they are an advance on that which defined an economic historian as one who wrote as little history as possible for as much money as possible. Economic history asks economic questions – be they about the demand and supply of goods and services, about costs of production, levels of income, the distribution of wealth, the volume and direction of investment, or the structure of overseas trade – it inevitably deals with large numbers, with aggregates. A study of economic history is important because the historical economic phenomena to be examined in any given period have no existence independent of the social, political, cultural, religious and physical environment in which they occurred.

1.3 Why Study History of Economic Thought?

The economic process that evolves in any society is a complex matrix of individuals, organizations, rules, and relationships. This matrix is the product of perceptions, values, beliefs, knowledge and technology. The economic process is embedded in society and is related to all aspects of the culture. An understanding of the economic system and economic theory requires an awareness of the social, historical and philosophical context in which they are developed. It is possible to train individual economists to apply economic tools, such as benefit/cost analysis without an understanding the historical and philosophical context of the tools. It is also possible to train an individual to fire and clean a firearm without considering the context of its alternative uses. This is the difference between training and education, between knowledge and wisdom. Both are needed. The history of economic thought is a study of alternative perspectives and explanations of how the economic processes function. An important aspect of the study of economic thought is to identify the factors that encourage different perspectives of the economy. It is also important to trace the evolution of the tools used for analysis and understand how the different perspectives and conditions encourage the use of different tools. Mark Blaug writes: "The task of the historian of economic thought is to show how definite preconceptions lead to definite kinds of analysis and then to ask whether the analysis stands up when it is freed from its ideological foundation. It is doubtful whether Ricardo would have developed his theory of international trade without a strong animus against the landed classes; but this theory survives the removal of his prejudices."

An understanding of the different approaches to economics, the causes for those differences and how they have evolved over time provides a historical and philosophical context that encourages a more critical analysis of current economic tools and their applications. This critical approach has three advantages. In many countries, the study of the history of economic ideas was previously considered to be an indispensable part of the training of an economist. However this point of view has been losing ground for a number of years. Many contemporary economists take no interest in the history of their subject, and some are decidedly doubtful about the value of acquiring historical knowledge. There may be several explanations for this, but a main reason is probably that modern economist more than their predecessors regard economics as a cumulative science in which new research and new insights are based on existing knowledge that is constantly being extended and improved. In a cumulative science, therefore, new insights will always tend to make the views of earlier scientists dated and erroneous. Some economists even argue that the history of thought was only about "the wrong opinions of dead men."

However the study of history of economic thought is still relevant because when you study a subject you should certainly have knowledge of how the subject evolved to its present form. A study of history of economic thought helps better understanding of the internal dynamics of the subject. The concepts and theories that today's students encounter in the syllabus are the results of the work of earlier generations of economists. The understanding that economics has developed as a continuous process which continues today is in itself an inspiration for those who wish to attempt to gain a better knowledge of the subject and perhaps even contribute to its further development. In addition, it could even be the case that the study of the older literature may encourage new research by the discovery and reconsideration of problems and fruitful insights that have been neglected in contemporary work. Following are some reasons why it might reasonable to use some time and effort getting to know the history of economic thought.

- ✰ It is exciting to explore the past. Anyone with some familiarity with modern economics should find it interesting to read about the thinkers and theories of the past. The opinions of dead men may be fascinating to study even if one believes them to be wrong. For example, Einstein's discoveries did not turn Newton into an irrelevant character in history; in a similar manner, Paul Samuelson and other twentieth-century economists did not make the life and work of Adam Smith a subject of no relevance and interest.
- ✰ There is great value in understanding the origin of a science, especially one like economics, whose scope and nature have been under dispute. It

gives a better concept of the relationship among sciences. This will help the thinker looking toward the proper application of economic principles.

- A study of the History of Economic Thought help to gain a clear understanding of the position of economics as a distinct member of a group of social sciences: ethics, jurisprudence, philosophy, sociology, etc.
- Some knowledge of the history of thought helps anyone to understand some phrases / concepts that are widely used even in mainstream media. For example, in general writing in newspapers and other media we often see excessive use of economic terms like (Adam Smith's) 'invisible hand,' while discussing about a capitalist country. Similarly we see 'Keynesian policies' while referring to recession. We read of 'laissez faire' while reading about LPG policies.
- Some familiarity with the history of thought contributes to a better understanding of the fact that the discipline of economics is in a permanent process of change and development, thereby leading to a better understanding of the nature of economic research. The history of thought makes one realize that economic science has always progressed through the efforts of people who have seen that it contains deficiencies and errors.
- In a time of economic crisis like we have now, a reflection of the roots of economic theory and methods prevents us from following the wrong path. For this study of history of economic thought is essential.
- A study of it helps to employ methods of literary interpretation in surveying earlier attempts on a theory.
- The history of economic thought displays a progressive rise to higher and higher levels of understanding of economic reality (as per the proponents of the cumulative view). They argue that such a study help in the creation of scholarship, 'scholarship', defined as 'the pursuit of broad and exact knowledge of the history of the working of the human mind as revealed in written Records'.
- John Maynard Keynes believed in the importance of studying and applying the wisdom of earlier economists. His works are studded with references to earlier economists.
- The study history of economic thought exposes one to the philosophical foundations of economics which is essential for a proper understanding of theories (micro, macro etc.).
- Further a study of history of economic thought helps to learn:
  - I. The intellectual heritage and a critical posture in dealing with texts
  - II. Principles of economics
  - III. From the classical works that have withstood the test of time
  - IV. From the masters
  - V. Economics as a history of economists
  - VI. To receive new insights for current research

VII. To understand the "affiliation of ideas," what succeeds, and how, and why
VIII. Guidance when the science undergoes revolutionary change
IX. Epistemological argument
X. Study of the competition of ideas
XI. Over time / across cultures
XII. Between schools
XIII. Concerning cyclical developments
XIV. With respect to different factor markets
XV. Preserving the stock of economic knowledge

1.4. Schools of Economic Thought

Historians categorize economic thought into 'periods' and 'schools'. This categorization is helpful for the purpose of exposition.

1. Classical School

The Classical School, which is regarded as the first school of economic thought, is associated with the 18th Century Scottish economist Adam Smith, and those British economists that followed, such as Robert Malthus and David Ricardo.

The main idea of the Classical school was that markets work best when they are left alone, and that there is nothing but the smallest role for government. The approach is firmly one of laissez-faire and a strong belief in the efficiency of free markets to generate economic development. Markets should be left to work because the price mechanism acts as a powerful 'invisible hand' to allocate resources to where they are best employed. In terms of explaining value, the focus of classical thinking was that it was determined mainly by scarcity and costs of production. In terms of the macro-economy, the Classical economists assumed that the economy would always return to full-employment level of real output through an automatically self-adjustment mechanism. It is widely recognized that the Classical period lasted until 1870.

2. Neoclassical

The neoclassical school of economic thought is a wide ranging school of ideas from which modern economic theory evolved. The method is clearly scientific, with assumptions, and hypothesis and attempts to derived general rules or principles about the behaviour of firms and consumers. For example, neoclassical economics assumes that economic agents are rational in their behaviour, and that consumers look to maximize utility and firms look to maximize profits. The contrasting objective of maximizing utility and profits forms the basis of demand and supply theory. Another important contribution of neoclassical economics was a focus on marginal values, such as marginal cost and marginal utility. Neoclassical economics is associated with the work of William Jevons, Carl Menger and Leon Walras.

3. New Classical

The New Classical School is built largely on the Neoclassical School. The New Classical School emphasizes the importance of microeconomics and models based on that behaviour. New Classical economists assume that all agents try to maximize their utility and have rational expectations. They also believe that the market clears at all times. New Classical economists believe that unemployment is largely voluntary and that discretionary fiscal policy is destabilizing, while inflation can be controlled with monetary policy.

4. Keynesian Economics

Keynesian economists broadly follow the main macro-economic ideas of British economist John Maynard Keynes. Keynes is widely regarded as the most important economist of the 20th Century, despite falling out of favour during the 1970s and 1980s following the rise of new classical economics. In essence, Keynesian economists are sceptical that, if left alone, free markets will inevitably move towards full employment equilibrium. They Keynesian approach is interventionist, coming from a belief that the self-interest which governs micro-economic behaviour does not always lead to long run macroeconomic development or short run macro-economic stability. Keynesian economics is essentially a theory of aggregate demand, and how best to manipulate it through macro-economic policy.

5. Monetarism

Monetarism is sometimes also referred to as the Chicago School (of economic thought).Monetarism is most widely associated with Milton Freidman and supports primarily a free market economy. Monetarist economists believe that the role of government is to control inflation by controlling the money supply. Monetarists believe that markets are typically clear and that participants have rational expectations. Monetarists reject the Keynesian notion that governments can "manage" demand and that attempts to do so are destabilizing and likely to lead to inflation.

6. Austrian School

The Austrian School is an older school of economics that is seeing some resurgence in popularity. Austrian school economists believe that human behaviour is too idiosyncratic to model accurately with mathematics and that minimal government intervention is best. The Austrian school has contributed useful theories and explanations on the business cycle, implications of capital intensity, and the importance of time and opportunity costs in determining consumption and value.

1.5 Ancient Economic Thought

For most of history, economics did not have a separate identity apart from social thought in general. Even as late as the eighteenth century, Adam Smith viewed economics as a subset of jurisprudence. Economics attained its distinctive identity when it came to be identified with a self regulating market process, and the discovery of the market as a self-regulating process was an eighteenth-century phenomenon. However, the seeds of economic analysis were sown long before, in ancient Greece, the cradle of Western civilization.

1. Greek Economic Thought

The theory of economics as a separate science never developed in Greece. The consideration of economic problems was incidental to the pursuit of politics and ethics. In so far as Greek thinkers treated such subjects, their theories reflect the comparative simplicity of their economic environment. Without prejudging the issue as to the actual extent of capitalism in ancient Athens, we need only to think away the vast international scope of our modern commercial problems, our giant manufacturing plants with their steam and electric power, our enormous wealth and its extreme concentration, the untold complexity of modern business and finance, the vast territorial expanse of modern nations, almost all our luxuries and commonplace comforts, to begin to appreciate something of this ancient simplicity. However, as a direct result of this limitation, the Greeks were led to deal with their problems more in terms of men than in terms of things, and thus their economic vision was sometimes clearer and truer than our own. Aristotle struck the keynote in Greek economic thought in stating that the primary interest of economy is human beings rather than inanimate property.

The economic ideas of Greek thinkers were not arrived at as a result of a purposeful study of the problems of material wealth. All economic relations were considered primarily from the standpoint of ethics and state welfare. The citizen was not regarded as a producer, but only as a possessor of wealth. The best economists are now insisting more and more on the Greek idea that economic problems must be considered from the standpoint of the whole man as a citizen in society. Modern political economy has placed man and not wealth in the foreground, and subordinated everything to his true welfare. Love, generosity, nobility of character, self-sacrifice, and all that is best and truest in our nature have their place in economic life. The Greeks felt that science which deals with wealth, so far from being a 'gospel of Mammon,' necessarily begins and ends in the study of man.

2. Economic Ideas of Plato

Plato (428 BC – 348BC) was a Greek philosopher of ancient times. He was also a mathematician, student of Socrates, writer of philosophical dialogues, and founder of the Academy in Athens, the first institution of higher learning in the Western world. Along with his mentor, Socrates, and his student, Aristotle, Plato helped to lay the foundations of Western philosophy and science. Plato's definition of economics, as suggested by one of the most recent historians of economic thought," could easily be accepted by many a modern scholar: "Economics is the science which deals with the satisfaction of human wants through exchange, seeking so to regulate the industries of the state as to make its citizens good and happy, and so to promote the highest well-being of the whole." The contention of the Socratics, that all economic operations must finally root in the moral, that all economic problems are moral problems, and that the province of economics is human welfare, is thus a dominant twentieth-century idea.

Plato's Theory of the Ideal State

In his most celebrated book the Republic, Plato gives the theory of an ideal state. As far as a state is concerned, Plato gives ideas about how to build an Ideal commonwealth, who should be the rulers of the Ideal state and how to achieve justice in the Ideal state. Plato finds the state as the more suitable place to discuss about the morality than an individual, because everything is easier to see in the large than in the small. A state, says Plato, is a man 'writ' large against the sky. The elements that make up a city correspond to the elements that constitute the individual human soul. The justice of the city is the same as it is for the individual. For Plato, there is not one morality for the individuals and another for the state. Like the tripartite individual human soul every state has three parts which are its three classes. The elements that constitute the human soul are as follows:

- Bodily appetite,
- Spirited elements,
- Reason. Like the tripartite individual human soul, every state has three parts such as:
  1. Producer class,
  2. Military class,
  3. Ruling class.

Plato finds the origin of the state in the various needs of people. Nobody is self-sufficient. So, to meet the various needs men created the political institution. To Plato, in the beginning there was only one class namely the producing class. Then emerged the guardian class. From the guardian class emerged the ruling class. In a state the producer class will consist of those people to whom the bodily appetites are dominant and who live for money. The producer class is made up of farmer, blacksmiths, fishermen, carpenters ashore –makers, weavers, labourers, merchants, retailers and bankers. The life of the producer class is much easier than the life of the rulers or the guardians. The life of the produce class follows the old familiar patterns of home and property, family and children, work, rest and recreation. By nature the producers have money. Each member of the producer class will be educated by being taught a trade or a profession – farming, banking, carpentry- according to his or her capabilities and to the needs of the society, both of which will be determined by the guardians. The military class will be drawn from that type of men to whom the spirited element is dominant and who live for success in aggressive and courageous acts. The members of the ruling class will be drawn from that type of man to whom reason is dominant and who lives only for truth. A state should be ruled only by the elite group of the most rational. In the ideal state each of these three classes will perform a vital function on behalf of the organic totality of the state. Selection of the ruling class Plato gives most emphasis on the selection of the ruling class. The selection of the ruling class is from all classes by natural intellectual capacity. Women as well as men possess the natural capacity of intelligence to become members of the ruling class. Plato proposes that an ideal state will be governed by a person who is highly educated, has passion for truth and has achieved the greatest wisdom of knowledge of the good. The ruler of this ideal state is called the Philosopher king. The Philosopher king has several important functions to perform. The rulers, said Plato, should be the one who has been fully educated, one who has come to understand the difference between the visible world and the invisible world, between the realm of opinion and the realm of knowledge, between appearance and reality. The Philosopher king is one whose education, in short, has led him up step by step through the ascending degrees of knowledge of the divided line until at last he has a knowledge of the good. To reach this point, the Philosopher King will have progressed through many stages of education. By the time he is eighteen years old, he will have had training in literature, music and elementary mathematics. His literature would be censored. Music also would be prescribed so that seduction music would be replaced by a more wholesome, martial meter. For the next few years there would be extensive physical and military training. At the age twenty a few would be selected to peruse an advanced course in mathematics. At age thirty, a five year course in dialectic and moral philosophy would begin. The next fifteen years would be spent gathering practical experience through public service. Finally, at age fifty, the ablest men would reach the highest level of knowledge, the vision of the good and would then be ready for the task of governing the state.

Both the ruling class and the military class are forbidden to possess any private property or any money. They must live, men and women like soldiers in barracks, with common meals and sleeping quarters. Their food, clothing and equipment will be provided by the producers. This food must be simple and restricted to moderate quantities. They are too have no family life, in order to avoid any conflict between family loyalties and their loyalty to the state. When they are at the physical prime of life, their sexual gratification is restricted to officially designated and infrequent occasions on which they are required to breed children to maintain the number of the guardian class. These occasions Plato calls sacred Marriage which are temporary unions for the sake of producing children.

Justice in the State

Like the human soul, the justice will be achieved in a state when each class fulfils their respective functions. Justice is a general virtue. It means that all parts are fulfilling their special functions. As the craftsmen embody the element of appetite, they will also reflect the virtue of temperance. Temperance is not limited to the craftsmen but applies to all the classes, for it indicates, when it is achieved, the willingness of the lower to be rulled by the higher. Still temperance applies in a special way to the craftsmen subordinate to the two higher levels. The guardians, who defend the state, manifest the virtue of courage. To assure the state that these guardians will always fulfill their functions. Special training and provision are made for them. Unlike the craftsmen, who marry and own property, the guardians will have both property and wives in common. Plato considered these arrangements essential if the guardians were to attain true courage, for courage means knowing what to fear and what not to fear. The only real object of fear for the guardian should be fear of moral evil. He must never fear poverty and privation, and for this reason mode of life should be isolated from possessions. Thus, in his Republic

Plato gives the theory of an ideal state. But later the theory of the ideal state was severely criticized by Aristotle.

Theory of Education

Plato supports state regulated system of education. According to Plato education does not mean the storing up of external knowledge but the bringing of the soul into proper environment for the development of the state. According to Plato a state can regulate crime by adopting a proper system of education. Plato gives more importance to education than any other Greek thinker.

Communist Social System

Plato propounded theory of communism of wives and property. Plato suggest communism for guardian and ruling class as he thinks both the community of property and the community of families tend them to make truly ruling class. Plato was of the opinion that the ruling elite class should not have any property of their own beyond what is absolutely necessary.

1.6 Economic Ideas of Aristotle

Aristotle (384 BC – 322 BC) was a Greek philosopher who was a student of Plato and teacher ofAlexander the Great. His writings cover many subjects, including physics, metaphysics, poetry, theatre,music, logic, rhetoric, linguistics, politics, government, ethics, biology, and zoology. Together with Plato and Socrates (Plato's teacher), Aristotle is one of the most important founding figures in Western philosophy. Aristotle's writings were the first to create a comprehensive system of Western philosophy, encompassing ethics, aesthetics, logic, science, politics, and metaphysics. The economic ideas of Aristotle are developed mainly in the following works: 'Politics', 'Nicomachean Ethics', 'Rhetoric', 'Economics' and 'Rhetoric to Alexander'. Aristotle analyses Economics according to ethical principles and examines it micro economically and macro economically. He based economics on needs, analyzed their nature and proceeded to isolate the economic goods by which economic needs are satisfied; he talks about production and the factors involved, the distribution of labour, the significance of the primary, the secondary and the tertiary sectors, and the stages in the development of the economy. He also examines the phenomenon of economy of an area, of economic development and prosperity on the basis of the financial policy. He also included the subjective perception of value, so that the influence of his intellectual work, as is shown, continues to appear up till the present time; thus he has influenced economic thought more than anyone else throughout History. The aim of Aristotle was the prosperity of the City-State along with its self-sufficiency sand the division of labour.

The nearest approach made by Greek philosophy to developing a distinct theory of economics came in discussing the elements of household management. Here a distinction was drawn between economics (oikonomik) and chrematistics (chrematistik); the former embraces chiefly wealth consumption in the satisfaction of wants, and the provision of such necessary and useful commodities as can be stored to meet those wants; the latter deals with wealth-getting, including money-making and exchange. Concerning the latter, Aristotle says, "And there is another element of a household, the so called art of money-making (or finance) which, according to some, is identical with household management, according to others, a principal part of it." It is the "natural" or "proper" branch of chrematistics alone which should be included in economics or household management. Closely connected with the preceding analysis is the distinction between the natural or proper and the unnatural or improper uses of a thing. "Of everything which we possess there are two uses: both belong to the thing as such, but not in the same manner, for one is the proper, the other the improper or secondary use of it. For example, a shoe is used for wear, and is used for exchange; both are uses of the shoe." This distinction rests upon Aristotle's notion of exchange, which, in its turn, is founded on the idea that there is a certain consumption which is sufficient for a proper life; for, when he says that retail trade is not a "natural" part of money-making, he adds that "had it been so, men would have ceased to exchange when they had enough." In other words, natural chrematistics concerns the satisfaction of natural or proper wants by "natural" or "proper" or "primary" uses.

1.6.1. Aristotle and Private Property

The ideas of Aristotle have had a tremendous impact on social and economic thought since the days of the Lyceum. Aristotle's greatest contribution is his recognition of the vital importance of private property. Aristotle denounced the communism of the ruling elite advocated by Plato. According to Aristotle, Plato's collectivist utopia runs counter to humanity multiplicity and the mutual advantage gained through market exchange. Plato himself recognized the importance of the division of labour. Plato has Socrates remark in The Republic that specialization occurs because we are not all alike; there is much diversity of natures among us which are adapted to different occupations. Aristotle outlined the common characteristics of private property that solidified his support:

- Private property is more productive and leads to progress.
- Conflict is inherent in communal property management.
- Private property is intrinsic to man's nature. The love of self, money, and property is tied to natural love of exclusive ownership.
- Private property has existed always and everywhere.
- Only private property allows for opportunity for moral action; to practice virtues of benevolence and philanthropy.

Aristotle's work on money was the backbone of medieval thinking about commerce. Marx's theory of economic value was based on it, and so was much of the economic analysis of money into the present century. In the past hundred years the interpretation of Aristotle's work on money has become chaotic. It is argued here that Aristotle does develop a coherent theory of economic value, wealth, exchange, and money. From an Aristotelian standpoint, ethics and economics are competitors over the same ground, as rival sources of reasons for decision-making in the public realm, and they cannot be reconciled. Aristotle also had a generally positive and accurate view of money despite his unfortunate comment that the lending of money at interest was unnatural. He correctly identified the growth of money as a catalyst for increased production and exchange. He sees that money as a medium of exchange is representative of general demand and thus holds all goods together. Aristotle goes on to explain that as everyone sells goods for money the problem of the double coincidence of wants is eliminated. No longer does each trader have to covet the other trader's goods directly.

Aristotle appreciates the fact that money represents human need or demand thus providing the motivation for exchange and which holds all things together. Demand is governed by the desirability of a good or use-value. Aristotle approaches a cogent analysis of the impact of different levels of supply on the value of a good. An echo of a marginal utility theory of value and its resolution of the paradox of value can also be discerned. Aristotle states that the quantity of a good reaches a saturation point where the use-value plummets and becomes insignificant. He points to the inverse effect that when a good becomes scarcer, it will become subjectively more valuable. In the Rhetoric, Aristotle states that what is rare is a greater good than what is plentiful. Thus gold is a better thing than iron, though less useful.

Although not a complete refutation of the paradox of value, much closer than many economists of the eighteenth century. Aristotle correctly identified the process by which the value of final products is imputed to the factors of production or means. In the Topics as well as in the Rhetoric, Aristotle states that the instruments of production derive their value from the instruments of action, the final products useful to man. The greater subjective value of a good, the greater the value of the means to arrive at that product. Aristotle touches on the marginal component in this imputation stating judge by means of an addition, and see if the addition of A to the same thing as B makes the whole more desirable than the addition of B. Aristotle correctly emphasizes the value of the loss rather than the addition of a good. That is the greater good whose contrary is the greater evil, and whose loss affects us more. Aristotle's analysis continued as he pointed out that a saw is more valuable than a sickle in the carpentry profession, yet it is not universally more valuable. The factors of production play an important role in the whole process. Aristotle also noticed that a good with many potential uses will be more desirable than a good with only one use. Aristotle provided clear insight into the economic theory of imputation and marginal productivity tackled by economists more than two thousand years later.

1.7. Compare and Contrast Aristotle's and Plato's Ideal State

As philosophers of the golden age of Greek philosophy, Plato and Aristotle have immensely contributed to political philosophy, aside other areas. In this write-up, we intend to evaluate the points of agreement and disagreement as regards the prescriptions on the ideal state by both these Socratic philosophers.

The Ideal State

Initially, a state is defined as a "territorial entity divided into government and subject; and claiming within its allotted area, supremacy over all other institutions". The word "ideal" simply means a "perfected standard". Hence, an ideal state must be a state that is based on a perfected standard. Plato and Aristotle both prescribed what these perfected standards on which the state be based, should be, according to their metaphysical orientation about man. Though, quite a number, a few points on which they both agree are given next;

1.7.1. Comparing Plato and Aristotle's Ideal State

a. For both, the end of the state is ethical; as justice is the basis for the ideal state. For Plato, the individual and the state are one, as they both have a tripartite nature of which justice is the result of a sound balance of these three parts. Aristotle asserts that the city-state (polis) comes into being for the sake of life, but exists for the sake of the good life.

b. Critics of Democracy – Both perceived democracy as the worst form of government. For Plato, democracy is the worst of all lawful (best) governments and the best of all lawless (worst) ones. For Aristotle, "a perverted polity degenerates into democracy (a rule by the mob) which is a bad form of government.

c. Education: A national concern – Plato prescribed that everyone must be given an equal opportunity in order to prove their mettle as regards where they belong in the social stratification. This is why he established the academy. For Aristotle, there should be laws guiding education as a national concern. For each is a part of the state and the care bestowed on each part, naturally tends towards the care of the whole. Aristotle, like Plato, also established a school "the lyceum".

d. Slavery and stratification of citizens -Both considered slaves as properties of their masters and justified the fact that some people are by nature slaves. The fact that both philosophers hail from aristocratic (wealthy) families could be the reason for this. Likewise, as regards the social classifications of citizens, for Plato it is the "guardians", "auxiliaries" and "artisans". For Aristotle it is the "rich", "middle class" and "poor",

e. Size of the ideal state - Both prescribed that the ideal state be small in size, in order to attain perfection easier. The Greek state (Athens) was relatively small. Also, it happened that the whole of Greece was divided into small city states, of which each had their own autonomous government and ruling system.

1.7.2. Contrasting Plato and Aristotle's Ideal State

Plato prescribed in his communism concerning the abolishment of private property, especially to the guardians and the auxiliaries. Plato says anything (property) that must be owned by them, must be owned collectively. This could be as a reason of their societal hierarchy and the tendency to abuse such privilege. Aristotle on the other hand criticized the abolition of private property, as he recognizes the need to own such even though the private possession of goods must be within certain limits. He therefore preached against the excessive accumulation of goods and advised citizens to use their limited possessions for the benefit of the common good. Plato's republic ideal state supports feminism, in that education for ruling as well as the job for ruling itself should be open to girls and women like the men. Thus, one's sex is generally irrelevant to ones qualifications for education or employment. Aristotle asserts that the women are normally subordinate to men, for the male is by nature superior and the female inferior. The men rules but the women are ruled (though not as slaves). Thus, Aristotle accepts the customary patriarchal subordination of women to men. Plato's ruling ideology has been summarized as the "rule of the best man" – the philosopher king who alone knows the ideal standards for the state. Also, ruling is a skill; as the best man must be trained to rule. Ruling is also an ideal. Aristotle's ruling ideology has been summarized as the "rule of the best laws" – a well ordered constitution which entails good governance. For him, although ruling is a skill and an ideal as well; it is also a science (although Aristotle understands politics as a normative or prescriptive discipline rather than as a "purely" empirical or descriptive inquiry).Plato proposes the abolishment of the family in his communism, as he says the guardians and the auxiliaries shall have no wife of their own, but in common. Children should be separated from their parents at birth and raised by the state. Thus, there will be more unity and fewer disharmonies. Aristotle disagrees and upheld that the family is the bedrock of the state and fundamental society established according to the law of nature to provide man's daily needs. He despised communism, in his words "everybody is inclined to neglect something which he expects another to fulfill; as in families many attendants are often less useful than a few".

2

The Analytical Contributions of Mercantilists to the Economic Theory

2.1. The Analytical Contributions of Mercantilists to the Economic Theory

(1) Concept of Nationalism

Mercantilism was never more than a means. The true end was political in character. That is creation of a strong state. The mercantilist emphasized on national strength and prosperity. The building up of a nation state was put in the fore front. Monetary and other economic devices were regarded merely as instruments to make the state strong. A state intervention was an essential part of mercantilist doctrine. Mercantilists believed that every aspect of life was subject to regulation by rulers for widening of market beyond medieval limit. Economically expansion of commerce and trade and politically success in war were their natural goals. This can be achieved only by a strong Nation built upon the spirit of nationality. For protecting the domestic trade and for expanding the foreign market a strong central government becomes an important element in mercantilist doctrine. So the mercantilists gave much importance for the creation of a strong Nation. Thus the mercantilists demanded a state strong enough to protect the trading interests and to break down the medieval barriers to commercial capitalism.

(2) Importance of Treasure

To the mercantilists the most important concern was the strength of the country. In considering the prosperity and strength of the country the true mercantilist had always at the back of his mind a comparative standard. To the mercantilists the strength of his country depends upon the stock of wealth which the nation possesses. By wealth they mean the stock of precious metals. According to the mercantilists the great and ultimate effect of trade is not wealth at large but particularly abundance of silver, gold and jewels, which are not perishable. Tile king who could support a large army and maintain a powerful navy was the king who had behind him the treasure necessary for this purpose. Hence the mercantilist attached great importance to treasure and bullion as the most useful and generally accepted form of wealth. The important slogans raised at the time were more "gold more gold, wealth more wealth"

(3) Foreign Trade and Balance of Trade

The mercantilist considered treasure (Gold, Silver, precious metals) as wealth par excellence. What are the means to be adopted for securing the desired treasure and wealth? Thomas Mun says "the ordinary means to increase our wealth and treasure is by foreign trade and this ought to be encouraged". Thus the mercantilist considered foreign trade as the only source for acquiring gold and silver. They believed that treasure gained by the balance of our foreign trade remained in the kingdom. In respect of its exports a country must receive payments in the form of silver or gold. On the other hand what it buys in the form of imports will have to be paid for. Export represent money coming into the country and import represents money going out. In order to increase the supply of Bullion in a country, it is necessary that there be a "favourable balance of trade" represented by excess of export, over import. So it is easy to conclude that all exports in the nature of goods are good and desirable whereas all imports are evil and damnable, it is thus a primary principle of typical mercantilists to maximize exports and to minimize imports. There was however, no uniformity of opinion among the mercantilist writers, so far as the concept of favourable balance of trade was, considered. Some of them advocated the balance of margin theory 'where every individual transaction must bring forth a gain to the state. While talking of a favourable balance of trade, most of the mercantilist did not signify whether they meant on overall favourable balance of trade or a favourable balance with each country. The mercantilists recognized the importance of invisible items of balance of trade such as freight income, insurance, diplomatic and military expenditure abroad. Thus the mercantilist supported a favourable balance of trade as the means to acquire more treasure.

(4) Industrial and Commercial Regulations

The mercantilists followed a number of policies in order to maximize the net gain from foreign trade. The main strategy is to increase the production of exportable commodities in the domestic economy and to reduce the import of articles from foreign countries. In order to maximize output in the domestic economy, the mercantilist suggested the following methods.

(a) Land resources may be fully utilized and fallow lands may be brought under cultivation (b) Employment opportunities should be increased, so that unemployed human resources can be utilized for increasing production of goods meant for export (c) Better method and technique of production should be introduced in the production process (d) In order to increase production and employment, 'the mercantilists advocated a system of low wage level. (e) Colonies can be utilized as source of raw materials to the mother country. Along with this a series of trade regulations and policies were framed by the mercantilists. These policies were aimed at export promotion and import restriction. The important among them were

- ☆ Restraints on the importation of foreign goods especially manufactured goods
- ☆ Encouragement to the export of manufactured products
- ☆ Restriction on the export of raw materials
- ☆ Restriction on the industries which interfere with other industries or trades of greater national importance.
- ☆ Ban on the export of bullion
- ☆ Establishment 'of banks and the promotion of use of credit instruments.

In order to execute all their schemes and programmes, the mercantilists looked to a benevolently paternal govt. which can interfere everywhere. There was nothing the govt. might not do; there was nothing the govt. ought not to do, if its activities help to promote the general wellbeing. So the mercantilists attached great importance to state as the centre of all economic activities. The control and regulation of the govt. are essential for the attainment of favourable balance of trade. The above list of proposals and govt. regulations reflect the protectionist philosophy prevailing in the mercantilist literature. Almost all kind of protectionist arguments are available in the mercantilist writing. The mercantilist protection argument can be summed up as follows:

(a) Protection may be granted to strategic industries. These industries which are strategic for the nation both in terms of their export contribution and also in terms of their contribution to economic development can be earmarked by the government of the country. (b) Protection may also be granted to the weak but promising industries which were at the- infant stage of growth. (c) Protection may also be granted to the defence industries (d) those industries should be given protection which help for the attainment of self-sufficiency and self-reliance for the country. (e) Protection may also be given to industries which are engaged in producing import substitutable and export promoting goods. From this, it appears that the mercantilists' writers were well balanced in their ideas on the different aspects of foreign trade.

In mercantilist writing an important role was assigned to money. The theoretical structure of mercantilist monetary system is based on three prepositions

(a) Wealth consists in money

(b) Money was not to be accumulated as treasure in the literal sense of stock of valuable and early saleable objects

(c) Money was desired to be used as a medium of exchange. Just as an individual saves out of his income, a country must spend less than its income, if it wants to increase its wealth. Thus they identified favourable balance of trade with the annual balance of income over consumption and money with capital. To the mercantilist wealth of a nation lay in its productive capacity. Money strengthens the productive capacity of a nation by means of it being in circulation. Mercantilists disfavoured the idea of accumulating money as treasure. Mercantilists believed that money would be utilized in production of more commodities; it will also create new demand for commodities through generation of employment. Thus he believed that increased money supply would result in increasing output and employment without allowing prices to' rise. Thus mercantilist maintained the doctrine that money stimulates trade.

2.1.1. Particular Economic Theories

(a) Value

Value was inherent in commodities and depend upon human needs. Value was regarded as a market phenomenon, which depends upon exchange. Mercantilists distinguished between two types of value. That is intrinsic value and extrinsic value. Intrinsic value is the power of the commodity to satisfy human wants. By extrinsic value, the mercantilists mean only the cost of production. William Petty clearly mentioned that value depends upon expenses of production. According to him market value was the extrinsic value of commodity which fluctuated according to a rise or fall in the supply and demand. Locke indicated the importance of labour and its contribution towards determination of value. He was the forerunner of labour theory of value which was developed later.

(b) Production

According to mercantilists, production was a process of application of human labour to resources and hence they advocated an increase of labour and resources. Mercantilists regarded agriculture as insignificant. They believed that foreign trade alone was the, roost productive. They made a distinction between productive and unproductive labour. The labour of a merchant, a manufacturer and a farmer was considered productive. Labour of professional persons like doctors, stock brokers and shop keepers was considered unproductive.

(c) On Taxation

The views of mercantilist on taxation were quite interesting. Mercantilists favoured a multiple tax system based on the basic principle of each should pay according to the benefits received from the state. Hobes held that man should be taxed according to the expenditure incurred by him, in his opinion, a tax on expenditure is the most equitable. Generally speaking the mercantilists favoured law customs, excise duties, a tax on interest etc. Their chief cannon of taxation was that of equity.

(d) On Population

Mercantilist favoured increasing population for maintaining military strength. It was also advocated for increase in productive capacity of the economy. They believed that cheap and abundant supply of labour would help in keeping the cost of production at lower levels. They encouraged family life and parenthood.

(e) On Interest

There was no uniformity of opinion regarding the interest concept. Thomas Mun, Locke and North favoured interest taking in money lending. According to them money lending provided the necessary capital to the poor merchants. Further it also led to gainful employment of the saving of widows, orphans and others. But the mercantilists were opposed to high rate of interest.

2.2. Important Mercantilist Writers

Sir William Petty (1623-1685)

Petty is regarded as the founder of political economy. He is remembered for his contribution in the field of statistical methods and economic theory He had a quantitative bend of mind. He was the first to develop a fact finding approach on economic enquiry. As a statistician Pettry confined himself only to the employment of quantitative data and used simple average as the statistical technology. His most famous work was the Treatise of taxes and contributions (1662).

Thomas Mun (1571-1641)

He was born in England in 1571. He was a merchant by profession. His famous book is England'sTreasure of Foreign Trade. Mun was responsible for shaping the trade policies of England during his time. He advocated the fundamental rule of International Trade. To him the rule of international trade should be "to sell more to strangers yearly than we consume of their in volume." He was an ardent believer of nationalism and a strong govt. This according to him would be realized only by accumulation of treasure. He suggested imposition of heavy import duties on goods meant for domestic consumption and moderate duties on export. He also recognized the importance of taxation.

David Hume (1711 -1n6)

John Locke studied the effect of a once - for - all change in money supply on the price level. Hume dynarnised Locke's argument by establishing a causal relationship between money supply and price level, Hume recognized the existence of proportionate relationship between money supply and price level, keeping volume of trade and velocity of money constant: He considered only the standard of value and medium of exchange function of money. He ignored the store of value function of money.

Richard Cantillon (1680 -1734)

Cantillon strictly followed Petty in his method and problems. Cantillon's method was uantitative and his problem was to discover a par between land and labour. Cantillon defined wealth as everything that gives satisfaction and labour and capital as the true source of wealth. Capital was regarded as a dependent factor produced jointly by land and revenue. In his discussion on value. Cantillion defined normal price of a commodity as 'a measure of the quantity and quality of land and labour entering into its production. Cantillon's most significant contribution was in the field of money. He analyzed the relationship between money supply and price level. To sum up Cantillon's fame lies in his analytical contribution to the theory of value and money and his description of the circular flow of national income.

2.3. Criticism of Mercantilism

Mercantilism was severely criticized. Advocates of laissez-faire argued that there was really no difference between domestic and foreign trade and that all trade was beneficial both to the trader and to the public. They also maintained that the amount of money or treasure that a state needed would be automatically adjusted and that money, like any other commodity, could exist in excess. They denied the idea that a nation could grow rich only at the expense of another and argued that trade was in reality a two-way street. Towards the end of 17th century, the mercantilist theories and practices were strongly opposed by many writers who championed the cause of individual liberty and agricultural productions. Even Petty, Locke and North of England criticized the state controlled foreign trade as against freedom of trade. But severe criticisms were levelled by the physiocrats and Adam Smith. The following criticisms were levelled against mercantilism by the opponents:

- The mercantilists exaggerated the importance of Commerce to the extent of depressing agriculture and other branches of human industry.
- Undue importance was attached to gold and silver.
- They were under erroneous belief that a favourable balance of trade alone would bring prosperity to the country.
- Their idea about value, utility capital and interest were vague and imperfect.
- They are narrow minded nationalists and not cosmopolitans. They could not conceive the ideas of mutually advantageous trade. However we cannot dismiss their ideas as useless or impractical. The idea of nationalism, self-sufficiency and economic strength were the outcome of their policies. The mercantilist policy proved successful in France, England, Holland and Germany who were competing for colonial supremacy.

2.3.1. Physiocracy

The Physiocrats were a group of economists who believed that the wealth of nations was derived solely from agriculture. Their theories originated in France and were most popular during the second half of the 18th century. Physiocracy was perhaps the first well developed theory of economics. They called themselves *économistes* (economists) but are generally referred to as Physiocrats in order to distinguish them from the many schools of economic thought that followed them. Physiocrat is derived from the Greek for "Government of Nature".The principles of Physiocracy were first put forward by Richard Cantillon, an Irish banker living in France, in his 1756 publication *Essai sur la nature du commerce en géneral* (Essay on the Nature of Commerce in General). The ideas were later developed by thinkers such as François Quesnay and Jean Claude Marie Vincent de Gournay into a more systematic body of thought held by a united group of thinkers. The Physiocrats saw the true wealth of a nation as determined by the surplus of agricultural production over and above that needed to support agriculture (by feeding farm labourers and so forth). Other forms of economic activity, such as manufacturing, were viewed as taking this surplus agricultural production and transforming it into new products, by using the surplus agricultural production to feed the workers who produced the extra goods. While these manufacturers and other non agricultural workers may be useful, they were seen as 'sterile' in that their income derives ultimately not from their own work, but from the surplus production of the agricultural sector. The Physiocrats strongly opposed mercantilism, which emphasized trade of goods between countries, as they pictured the peasant society as the economic foundation of a nation's wealth. The Physiocrats enjoyed some support from the French monarchy and frequently met at Versailles. AdamSmith, who visited France as a tutor and mentor to the Earl of Buccleigh's son's Grand Tour, was heavily influenced by the ideas of the Physiocrats, and Karl Marx cites them as a reference in Das Kapital; they popularized the modern version of the labour theory of value.

Physiocracy is a school of thought founded by François Quesnay (1694-1774), a court physician to King Louis the 15th. At one point in time Physiocracy constituted a sort of religious movement that attracted a number of outstanding and extremely fervent believers, and exerted no small influence on real politics. The history of the Physiocratic movement is thought to have begun in 1757, when Quesnay met Mirabeau the elder (1715-89), and come to an end in 1776, with the fall of Turgot (1727-81). The actualmembers of the Physiocratic school referred to themselves not as Physiocrats but as économistes. Theterm "physiocracy" apparently came into general use after having first appeared in 1767, with theappearance of a collection of Quesnay's works published by Pierre du Pont under the title Physiocratie,ou Constitution Naturelle du Gouvernement le Plus Avantageux au Genre Humain. The term is of coursea combination of "physio" (nature) and "cracy" (rule), thus meaning the "rule of nature." This expresses the school's fundamental idea that there is a natural order, as opposed to artificial systems, and that the mission of scholarship and politics being to understand this natural order and bring it into existence, thereby bringing about this rule of nature. Physiocracy (from the Greek for "Government of Nature") is an economic theory developed by the Physiocrats, a group of economists who believed that the wealth of nations was derived solely from the value of "land agriculture" or "land development." Their theories originated in France and were most popular during the second half of the 18th century. Physiocracy is perhaps the first well-developed theory of economics.

The Physiocrats, or the "Economists" as they called themselves, were the first school of economicthought. They represented a reaction against the policies of Jean Baptiste Colbert [1619-1683]. Colbertwas served as a minister in the Court of Louis XIV. Colbert advocated strict regulation of commerce,protective tariffs and is regarded as a archetypical "Mercantilist." There were a number of writers whobegan to question the mercantilist policies of Colbert by the early 1700s (examples are PierreBoisguillebert [1646-1714], Seigneur de Vauban [1633-1707] and later Richard Cantillon [1680-1734]),however it was François Quesnay [1694-1774] who provided the basic structure of the Physiocratic system in the late 1750's. The Physiocrats represented an "alliance of persons, a community of ideas, and acknowledged authority and a combination in purpose, which banded then into a society apart."They held in common the idea that all things are part of an interconnected system that is rational and comprehensible to the human mind. Physiocracy was a revolt against mercantilism in France. This was due to the deteriorating economic and social conditions at that time which were chiefly due to mercantilist policies. The various factors responsible for the growth of physiocracy in France are enumerated below.

France was experiencing an absolute monarchy but without its potential benevolence. The life of the political administrators, from the king downwards, was very luxurious and corrupt. Public expenditure was very extravagant and wasteful. The result was an increasing indebtedness of the government and the need to levy extra taxes to finance the public expenditures.

The deteriorating economic conditions provided a new ground for new ideas and a welcome change. In France, during the reigns of Louis, taxation was unduly heavy, unjust and inequitable. The nobility and clergy owned about two-thirds of the country's land, but they were hardly paying any taxes while the poor peasants were being crushed under all sorts of levies in addition to the extortionist land rents. The manner of tax collection was also highly deplorable. Obviously the farmers were hardly left with any surplus which they could use for improvement of land or for improving their own consumption standards. The peasants also had to bear the burden of providing services to the feudal lords. There were many other oppressive taxes like the salt tax, poll tax, the tithe etc. which were equally burdensome for the poor sections of the community. The tortured life led by the farmers gave an impetus to the physiocratic policies and doctrines. Mercantilist policies had outlived their importance. Its policies were severely criticized. People were looking for an alternative system. In England agriculture was being revolutionized with the introduction of large scale farming. The achievements of this revolution reached the ears of the French people through the writings of Gourney, Mirabeau and Montesquieu. Agriculture in France was in a state of stagnation compared with its increasing usefulness and profitability in England. Colbertism was partly responsible for this. Industrial development was taking place in France at the cost of agriculture. Investments were diverted from agriculture to manufacture. The value of agricultural produce fell on account of restricted markets. Lower prices of agricultural products prevented capital accumulation in agriculture. Agriculture suffered due to lack of capital and initiative. In France a group of ambitious agriculturists and landlords started seriously thinking of the development of agriculture in France. They did not tolerate the suppression of French agriculture by king. One such ambitious landlord was Quesnay. The on-going Agricultural

Revolution in England influenced very much these ambitious French landowners. They thought that similar revolutions can be brought about in France.

According to Prof. Haney, there were great subjective factors at work for change and progress discarding mercantile policies. During the period of Louis 14, the people could not criticise his policies. After his death the people of France got sufficient liberty to express their opinion. This was the beginning of breaking away from the established policies, politics and religion. France witnessed an extreme form of inequality between different classes of people and between different sectors of the economy. There were both privileged and unprivileged classes. The common people were underprivileged who wanted an escape from this unwanted situation. Thinkers concerned with the ills of the society were trying to figure out and convey suggestions for the reformation of the system. There were analytical discussions and explorations regarding the ill effects of the existing system and the type of the ideal system which should replace. They were all agreed that the ideal system is physiocracy.

As Haney observes physiocracy was the revolt of French people against rnercantilistn. By the middle of the 18th century, the doctrine of mercal1tilism lost its practical usefulness and applicability. In Thomas Mun's own country - England - the doctrine was much discredited. Mercantilist policies gave undue emphasis to trade. In England an agricultural revolution was taking place due to large scale farming. The advantages of the revolution was brought to the knowledge of the French by writers like Mirabeau, and Mantesquiea, The works of several English writers were translated into French. This provided an opportunity to the French thinkers' to understand the progress of Great Britain, which depended to a great extend upon agricultural activities. The people were in search for an alternative system. They were looking for a new system of socio-economic order which could substitute mercantilism. This system was provided by physiocracy. According to Haney, there were great subjective forces at work for change and work. During the period of Louis XIV people could not criticize his policies because of his tyrannical rule. After the death of Louis XIV, the people of France got sufficient liberty to express their opinion, People began to break away from the established policies, politics and religion. People wanted something more rational, simpler and more practical. Above all they began to demand something which is based on natural justice. Thus naturalism came to the forefront of human action and thought. This demand was satisfied by the new system of physiocracy which was based on the' rule of nature.

2.4. The Basic Principles of Physiocracy

The following are the fundamental principles and policies of physiocracy.

- Agriculture is the only productive occupation.
- Industry and trade are sterile occupations.
- Agriculture alone produces net product.
- There is a natural order which makes life happy and meaningful.
- There is harmony among all classes of people.

- The individual should get maximum liberty.
- State action should be limited to the minimum.
- Trade is a necessary evil, and there should be free trade.
- Value depends on utility. Wealth has value. Value and price are the same things.
- The wage level is at the subsistence level.
- There is interdependence in the economic system.
- Real wealth lies in tangible and consumable goods.
- Private initiative must be encouraged.
- Distribution of products is very essential.
- Money is a medium of exchange.
- All that is bought is sold and all that is sold is bought.
- Rent is a perfectly legitimate income of the landlords.
- There should be a single and direct tax on land, as it is the only productive source.
- Private property is essential.
- There is the possibility of overpopulation on land.

2.5. Physiocratic Doctrines

The Economic Doctrines of the physiocrats can be conveniently classified under the following Four heads:

(1) Natural Order or Natural Philosophy

Natural order is the symbol of physiocratic system. Dupontde- Nemours defined physiocracy as "the science of natural order". Natural order, according to physiocrats, was an ideal order of things guided by the laws of nature, which has been ordained by god for the happiness of mankind. 'It was really the providential order. It was eternal and unchangeable. The' physiocrats conceived of an ideal order of things and a perfect arrangement of institutions guided by the laws of nature. Natural laws govern the activities of its members. The problem is to discover and obey these laws. There are two types of natural laws. Natural physical laws and natural moral laws. Physical laws govern physical universe. Natural moral laws are the laws of human actions. They are laws which govern human behaviour. They include rules of prudent individual conduct. It is the rule of justice, to be followed by individuals while dealing among themselves. According to physiocrats, natural order was an ideal order of things created by god for the maximization of human happiness. The man made social orders were artificial in nature. So the physiocrats wanted the people to do away with artificial manmade rules. Physiocrats believed that people suffer from social evils because the social order did not confirm to the natural order. In their opinion, the present organization of state should be suitably modified to bring the life, behaviour and ideas of the people within the orbit of the natural order.

(2) The Net Product

The concept of net product was the direct outcome of the philosophy of natural order of the physiocrats. The mercantilist has maintained that the source of wealth lay in foreign trade and that it consisted of precious metals. The physiocrats differed from the mercantilists on both these ideas. Physiocrats considered agriculture as the only source of wealth. To the physiocrats the only productive wealth was agriculture. To them the other occupations, other than agriculture were unproductive and sterile. In agriculture nature labours along with man. By the gift of nature (fertility) agriculture produces more than what the farmers consumes. This surplus production in the agriculture sector is called as "Net product". This surplus is obtained by subtracting the input (the amount of wealth destroyed in the production of new wealth) from the amount of new wealth produced. This surplus is used to nourish the other classes of people. It means that agriculture not only satisfied the needs of those engaged in it, but also of others engaged in trade, manufacture, and other professions. Thus the surplus can be taken as the difference between the quantum of production and the quantum utilized in the process of production. Physiocrats considered commerce and manufacture as unproductive. To the physiocrats production means creation of surplus. They therefore thought, only that industry to be productive because it increases the wealth of the nation. In Industry merely raw materials are transferred into finished products. In commerce and trade a mere transfer of wealth is taking place. In agriculture alone a surplus is generated. So labour which was applied anywhere except to land was sterile because man is not a creator. Thus the physiocrats believed that agriculture alone was productive. As a result the net product in the economy was possible and civilization had become a fact. Since agriculture was the sole source of wealth, it should be promoted, interest in it should be increased and hindrance from its way should be removed.

(3) The Circulation of Wealth

Based upon the concept of Net Product, the physiocrats advanced the theory of circulation of wealth. That is distribution of net product. Physiocrats were the first to attempt to analyse the problem of distribution. The credit for putting the idea in a systematic manner goes to Francois Quesney. He skilfully and graphically analysed the concept of circulation of wealth, drawing 'analogies from the circulation of blood in the' body. Being a doctor; he was able to describe it very eloquently. The circulation of net product in the words of Turqot constituted the very life blood of body politics, just as the circulation of blood of the physical. (François Quesnay's Tableau Économique is discussed in detail later).

(4) Individualism and Laissez Faire

The Physiocrats, especially Turgot, believed that self-interest was the motivating reason for each segment of the economy to play its role. Each individual was best suited to determine what goods he wanted and what work would provide him with what he wanted out of life. While a person might labour for the benefit of others, he will work harder for the benefit of himself; however, each person's needs are being supplied by many other people. The system works best when there is a complementary relationship between one person's needs and another person's desires, and trade restrictions place an unnatural barrier to achieving one's goals.

(5) Private Property

None of the theories concerning the value of land could work without strong legal support of private property. Combined with the strong sense of individualism, private property becomes a critical component of the workings of the Tableau.

(6) Diminishing Returns

Turgot was one of the first to recognize that "successive applications of the variable input will cause the product to grow, first at an increasing rate, later at a diminishing rate until it reaches a maximum". This was recognition that the productivity gains required to increase national wealth had an ultimate limit, and, therefore, wealth was not infinite.

(7) Investment Capital

Both Quesnay and Anne Robert Jacques Turgot, Baron de Laune recognized that capital was needed by farmers to start the production process, and both were proponents of using some of each year's profits to increase productivity. Capital was also needed to sustain the labourers while they produced their product. Turgot recognizes that there is opportunity cost and risk involved in using capital for something other than land ownership, and he promotes interest as serving a "strategic function in the economy."

(8) On Taxation

Physiocrats put forward a simple system of taxation. They advocated a single tax system, namely the land tax, which should be paid by the proprietary class directly to the govt. The theory of taxation was closely linked with the concept of net product. Some taxes were required for meeting the expense of the state 'for the maintenance of security, spread of education and establishment of public works. The only source which could be tapped was the net product and the only class who could pay taxes were the landed proprietors. So they advocated a single tax on land. This single tax was direct and hence cannot be evaded. It was simple to assess. The physiocratic scheme of taxation was tried in small regions in France, but did not succeed due to various reasons. However physiocratic ideas continued to prevail in the minds of statesmen: About 50% of the income budget of the constituent. Assembly (following the French Revolution) was from land taxes according to physiocratic ideas.

(9) On Value

Physiocrats had taken little interest in. the theory of value. According to Turgot value depended upon utility. But they did not regard value inherent in commodities; they also differentiated value in use and value in exchange. But they treated price and value as one and the same thing. Value according to physiocrats was not fixed but changed from time to time depending upon demand: Anyhow, the theory of value was not much important part of the physiocratic school.

(10) On Trade

As commodities of equal value were exchanged, trade and commerce were considered as unproductive. According to them trade did not produce any real wealth. Hence all commodity transition was sheer waste of time and energy. They wanted to counteract the evils of commerce by advocating complete freedom in the field of commerce.

(11) Functions of the State

The physiocrats were not anarchist. They suggested a state with minimum of civil laws which place no hurdles in the way of the realization of the natural order. To them the best state was the one in which there were the least number of laws. They advocated a state with minimum of legislation and maximum of authority. The function of the physiocratic sovereign were (a) to preserve natural order (b) To defend private property (c) To spread universal education (d) To undertake a programme of public work.

2.6. The Physiocratic School

The physiocratic were credited with the creation of the earliest school of thought in the propersense of the terms. They must be regarded as the real founders of the political economy. The membersof the physiocratic school published certain important books and articles which laid the foundations of physiocratic system. The spiritual leader of the physiocratic school was Boisguillebert, a lawyer who had contributed two books.

Francois Quetsnay (we have already seen his contribution)

Mirabeau

Mirabeau was a philosophical statesman. He took over the leadership of the physiocratic schoolafter the death of Quesnay. His important works were (1) The Friend of Mankind (2) The Theory of Taxation and (3) Rural philosophy. According Mirabeau Wealth is the fruit of earth or agriculture and labour of men alone has the power to increase it.

Robert Jaquis Turgot

Turgot after Quesnay was the most important member of the physiocratic school. He was the finance minister of Louis XIV. His important contribution - Reflections on the formations and Distribution of wealth published in 1766 was a treatise containing about 100 paragraphs wherein the superiority of agriculture was in full glory.

Dupont-de-Nemours

Nemours, the originator of the term physiocracy was an enthusiastic disciple of Dr. Quesnay. Heas chiefly a propagandist of the physiocratic doctrines and diu not make any original contribution to economic thought. He played a prominent part in French Revolution.

2.6.1. Critical Appraisal of Physiocrasy

The principal service of the physiocrats to modern political economy was not the discovery of any one of their doctrines, but their attempt to formulate a science of society out of materials already at hand. It was from this system as a base that Adam Smith set out to give a new impetus to the study of economic phenomena. Another important contribution consisted in calling attention to the weaknesses of the mercantile system. Laissez faire was a good doctrine for the eighteenth century because there was need of a reaction, but it was a mistake to set it up as a universal principle applicable under all conditions. The chief weakness in the physiocratic teaching lay in its theory of value. While agriculture brings forth the raw material of production, commerce and manufactures are equally productive of wealth. In a sense, the physiocrats recognized this, but they held that in producing this wealth the manufacturing and commercial classes use up an equivalent amount of value. This is a gratuitous assumption, but even if true, the same thing could be said of the so-called productive class. Moreover, if wages were governed by the "iron law" both in agriculture and in manufactures and commerce, as the physiocrats assume, the "net product" would be made up of wealth created by the commercial and manufacturing classes as well as by the agricultural class. The theory of the import unique or single tax rested upon the assumption that all incomes, except those of the proprietors, were at the existence minimum. Since this is not true, it is also not true that all taxes levied upon the other classes will ultimately be paid by the proprietors. The important criticisms leveled against physiocracy are as follows:

I. Their theory was drowned in normative statement. This is quite true of natural order concept.

II. The physiocrats failed to consider the labouring Class as a productive class. Moreover their contention that manufacturing class is sterile is also subject to severe criticism.

III. The physiocrats do not have a clear cut concept of value. They have confused value with utility. They held the view that value depend on utility.

IV. Their conception of landlord as partly productive class is more based upon political motive.

V. Physiocrats placed too much emphasis on agriculture and have neglected the non-agricultural sector.

VI. Hanex says that physiocratic doctrines are full of negative attitudes.

Anyhow physiocracy occupied an important place in the history of economic thoughts. They are notable in the history of Economic thought because of their constructive and positive contribution for the development of the science of Economics.

3

Classical, Socialist and Marxian School

3.1 Classical, Socialist and Marxian School

A. The Classical School

Classical economics refers to work done by a group of economists in the eighteenth and nineteenth centuries. They developed theories about the way markets and market economies work. The study was primarily concerned with the dynamics of economic growth. It stressed economic freedom and promoted ideas such as laissez-faire and free competition. Economic thought until the late 1800's. Adam Smith's Wealth of Nations, published in 1776 can be used as the formal beginning of Classical Economics but it actually it evolved over a period of time and was influenced by Mercantilist doctrines, Physiocracy, the enlightenment, classical liberalism and the early stages of the industrial revolution. Classical economics as the predominant school of mainstream economics ends with the 'Marginalist Revolution' and the rise of Neoclassical Economics in the late 1800's. In the 1870's William Stanley Jevons' and Carl Menger's concept of marginal utility and Leon Walras' general equilibrium theory provided the foundations. Henry Sidgwick, F.Y. Edgeworth, Vilfredo Pareto and Alfred Marshall provided the tools for Neoclassical economics. Neoclassical economics is an extension of Classical economics but, the focus of the questions changed as well as the tools of analysis. In spite of the dominance of Neoclassical thought, Classical Economics has persisted and influences modern economics, particularly the 'New Classical Economics.' The belief in the efficacy of a 'free market' is central to both classical and neoclassical ideology.

Famous economists of this school of thought included Adam Smith, David Ricardo, Thomas Malthus and John Stuart Mill.While Adam Smith would be regarded as the originator and leader of the school, David Ricardo should be credited with establishing the form and methods of the school. The debates betweenThomas Malthus and David Ricardo about policy issues such as the 'Corn Laws' and the 'Poor Laws' contributed to the focus and form of the school. Smith was concerned about the nature of economic growth. Malthus, Ricardo and other classical economists were concerned about the question of' distribution.' One important debate among classical economists was whether there was or wasn't a' surplus' or 'glut.' Jean Baptiste Say and Malthus were the two major protagonists in the question about the existence of a surplus and its effects on a market economy.

1. Adam Smith

Adam Smith was born in Kirkcaldy, Fife, Scotland. The exact date of his birth is unknown; however, he was baptized on June 5, 1723. Smith was the Scottish philosopher who became famous for his book, 'The Wealth of Nations' has written in 1776, which had a profound influence on modern economics and concepts of individual freedom. In 1751, Smith was appointed professor of logic at Glasgow University, transferring in 1752 to the chair of moral philosophy. His lectures covered the field of ethics, rhetoric, jurisprudence and political economy, or 'police and revenue.' In 1759 he published his Theory of Moral Sentiments, embodying some of his Glasgow lectures. This work was about those standards of ethical conduct that hold society together, with emphasis on the general harmony of human motives and activities under a beneficent Providence. Smith moved to London in 1776, where he published An Inquiry into the Nature and Causes of the Wealth of Nations, which examined in detail the consequences of economic freedom. It covered such concepts as the role of self-interest, the division of labour, the function of markets, and the international implications of a laissez-faire economy. 'Wealth of Nations' established economics as an autonomous subject and launched the economic doctrine of free enterprise. Smith laid the intellectual framework that explained the free market and still holds true today. He is most often recognized for the expression 'the invisible hand,' which he used to demonstrate how self-interest guides the most efficient use of resources in a nation's economy, with public welfare coming as a by-product. To underscore his laissez-faire convictions, Smith argued that state and personal efforts, to promote social good are ineffectual compared to unbridled market forces. In 1778, he was appointed to a post of commissioner of customs in Edinburgh, Scotland. He died there on July 17, 1790, after an illness. At the end it was discovered that Smith had devoted a considerable part of his income to numerous secret acts of charity. His Important works are: 'The Theory of Moral Sentiments' (1759).

1.1 Invisible Hand Theory

It is a term coined by Adam Smith in his 1776 book 'An Inquiry into the Nature and Causes of the Wealth of Nations'. One of the greatest contributions of Adam Smith was the invisible hand theory. He said that if the government doesn't do anything, there's a controlling factor of people themselves who can guide markets. I believe that the government should be responsible in defining the property rights, to set up honest courts, to impose minor taxes and to compensate for well defined 'market failures' If I sell candies for 1 peso each and Christian sells them for 2 pesos for 3 pieces, he will get all the business making me lose mine so in order to compensate for my loss I should be forced to lower my price as to stay alive in the business. I am guided by an invisible hand which is my self-interest to gain profit or as Adam Smith would say everyman for himself. The theory of the Invisible Hand states that if each consumer is allowed to choose freely what to buy and each producer is allowed to choose freely what to sell and how to produce it, the market will settle on a product distribution and prices that are beneficial to all the individual members of a community, and hence to the community as a whole. The reason for this is that self-interest drives actors to beneficial behaviour. Efficient methods of production are adopted to maximize profits. Low prices are charged to maximize revenue through gain in market share by undercutting competitors. Investors invest in those industries most urgently needed to maximize returns, and withdraw capital from those less efficient in creating value. Students prepare for the most needed (and therefore most remunerative) careers. All these effects take place dynamically and automatically. According to Wealth of Nations 'By preferring the support of domestic to that of foreign industry, he intends only his own security; and by directing that industry in such a manner as its produce may be of the greatest value, he intends only his own gain, and he is in this, as in many other cases, led by an invisible hand to promote an end which was no part of his intention. Nor is it always the worse for the society that it was not part of it. By pursuing his own interest he frequently promotes that of the society more effectually than when he really intends to promote it. I have never known much good done by those who affected to trade for the public good. It is an affectation, indeed, not very common among merchants, and very few words need be employed in dissuading them from it.' This is as how Adam Smith explained it that being led by an invisible hand is actually profitable in the sense that it uses the will of a person's self-interest which drives him to create more and better ideas to overcome the other competitors as long as he would be doing it in a legal way.

1.2. Naturalism and Optimism

Naturalism denotes the spontaneous and natural origin of economic institutions while Optimism refers to the belief that economic institutions so created were beneficial in their effects. Smith was emphatic that human actions guided by their selfish interests were bound to result in social good. To Smith spontaneity and beneficence went together. The presence of one implied the presence of the other. Every individual is motivated by self-interest. Every man, so to say, is the maximiser of gain and minimizer of loss. Therefore, every man undertakes that activity which appears to him to be the most profitable. The entire society is based on this conceptualization. Everybody wants to make his conditions better. Being guided by such a rationale, when people undertake different activities, the society gets many types of economic institutions. These economic institutions are spontaneous in character. But, although the institutions are primarily based on self-interest, they ultimately increase social welfare. It is not from the benevolence of butcher, the brewer or the baker that we expect our dinner, but from their regard to their on interest. Every man is naturally the best judge of his own interest and should, therefore, be allowed to pursue his own way. Adam Smith pointed out that in pursuing his own advantage, every individual is led by an invisible hand to promote an end which was no part of his intention. Smith believed that given the natural order, the interests of different economic units cannot be in conflict with one another. Their interests must be in harmony. This is ensured by the working of the Invisible hand or automatic adjustment mechanism. If perfect competition prevails, Smith things, individual interest and social interest would be compatible with each other. For the working of the natural order or naturalism, there must be liberty, free competition and laissez- faire. Adam Smith has given a number of examples of economic institutions which are spontaneous in origin and are beneficial in character. It is the instinct to barter and exchange one thing for the other that men resorted to specialization and division of labour. But division of labour gives rise to opulence and is hence beneficial. The institution of money arose as a spontaneous institution; but it removed the difficulties of barter system and facilitated trade, exchange and commerce and thus became useful. Adam Smith extended his principles of spontaneity to the supply of money and the distribution of gold and silver stocks between different countries of the world. There was a mechanism which automatically corrected any imbalance here. In the same say we may look at his contribution to the theory of capital accumulation where again the natural instincts of man lend a spontaneity of man lend a spontaneity to it and make it useful to the society. Another example of Smith's naturalism and spontaneity of economic institutions is his theory of demand and supply. Like Marshall, Smith pointed out that in the market, an excess of supply over demand would push down the price of the good under consideration and the price would rise in the opposite case. In this way a continuous process is always in operation whereby the balance between demand and supply gets restored. Smith also extends his 'spontaniety" in the case of population. He maintains that it is the variation in the wage rate that regulates the supply of labour through a variation in population.

Thus, in the Wealth of Nations the concept of 'naturalism' is combined with 'optimism'. Smith maintains that economic institutions are not only natural, but also beneficial. The institutions which have come into existence spontaneously through self-interest of individuals are natural and beneficial, as these institutions ensure progress towards wealth and prosperity.

However Smith's idea of Naturalism and Optimism is not free from criticism. His theory of optimism seems to be correct in the area of production. But in the area of distribution, there is hardly any scope for optimism and harmony of interest. Smith himself has shown that in the field of distribution there is the possibility of a class of interest between labour and capital. His theory of optimism is not based upon completely scientific foundation. Individual interest and social interest may not coincide with each other all the time. Similarly, the private sector and public sector do not have the same common interest and the same efficiency. The conclusion drawn by Adam Smith with respect to naturalism and optimism is based on insufficient inductive or empirical study. These are not based on any casual relation. It has been found that with the progress of capitalism, class division in society grows sharper and sharper. Smith is not right when he says that naturalism and optimism can be obtained without the help of the state. In fact, in many cases, state help will be necessary for obtaining optimism and naturalism.

1.3 Division of Labour

The main focus of Adam Smith's The Wealth of Nations lies in the concept of economic growth. Growth, according to Smith, is rooted in the increasing division of labour. This idea relates primarily to the specialization of the labour force, essentially the breaking down of large jobs into many tiny components. Under this regime each worker becomes an expert in one isolated area of production, thus increasing his efficiency. The fact that labourers do not have to switch tasks during the day further saves time and money. Of course, this is exactly what allowed Victorian factories to grow throughout the nineteenth century. Assembly line technology made it necessary for a worker to focus his or her attention on one small part of the production process. Surprisingly, Smith recognized the potential problems of this development. He pointed out that forcing individuals to perform mundane and repetitious tasks would lead to an ignorant, dissatisfied work force. For this reason he advanced the revolutionary belief that governments had an obligation to provide education to workers. This sprung from the hope that education could combat the deleterious effects of factory life. Division of labour also implies assigning each worker to the job that suits him best. Productive labour, to Smith, fulfils two important requirements. First, it must 'lead to the production of tangible objects.' Second, labour must 'create a surplus' which can be reinvested into production. Division of labour is the outcome of a tendency common to all men, the tendency to barter; and this tendency itself is spontaneously developed under the influence of personal interest, which acts simultaneously for the benefit of each and all.

Division of Labour' gives the example of a pin factory to explain the concept of division of labour. We will first see Smith's own words and then will interpret it in our own way. 'The effects of the division of labour, in the general business of society, will be more easily understood by considering in what manner it operates in some particular manufactures. It is commonly supposed to be carried furthest in some very trifling ones; not perhaps that it really is carried further in them than in others of more importance: but in those trifling manufactures which are destined to supply the small wants of but a small number of people, the whole number of workmen must necessarily be small; and those employed in every different branch of the work can often be collected into the same workhouse, and placed at once under the view of the spectator. In those great manufactures, on the contrary, which are destined to supply the great wants of the great body of the people, every different branch of the work employs so great a number of workmen that it is impossible to collect them all into the same workhouse. We can seldom see more, at one time, than those employed in one single branch. Though in such manufactures, therefore, the work may really be divided into a much greater number of parts than in those of a more trifling nature, the division is not near so obvious, and has accordingly been much less observed. To take an example, therefore, from a very trifling manufacture; but one in which the division of labour has been very often taken notice of, the trade of the pin-maker; a workman not educated to this business (which the division of

labour has rendered a distinct trade), nor acquainted with the use of the machinery employed in it (to the invention of which the same division of labour has probably given occasion), could scarce, perhaps, with his utmost industry, make one pin in a day, and certainly could not make twenty. But in the way in which this business is now carried on, not only the whole work is a peculiar trade, but it is divided into a number of branches, of which the greater part are likewise peculiar trades. One man draws out the wire, another straights it, a third cuts it, a fourth points it, a fifth grinds it at the top for receiving, the head; to make the head requires two or three distinct operations; to put it on is a peculiar business, to whiten the pins is another; it is even a trade by itself to put them into the paper; and the important business of making a pin is, in this manner, divided into about eighteen distinct operations, which, in some manufactories, are all performed by distinct hands, though in others the same man will sometimes perform two or three of them. I have seen a small manufactory of this kind where ten men only were employed, and where some of them consequently performed two or three distinct operations. But though they were very poor, and therefore but indifferently accommodated with the necessary machinery, they could, when they exerted themselves, make among them about twelve pounds of pins in a day. There are in a pound upwards of four thousand pins of a middling size. Those ten persons, therefore, could make among them upwards of forty-eight thousand pins in a day. Each person, therefore, making a tenth part of forty-eight thousand pins, might be considered as making four thousand eight hundred pins in a day. But if they had all wrought separately and independently, and without any of them having been educated to this peculiar business, they certainly could not each of them have made twenty, perhaps not one pin in a day; that is, certainly, not the two hundred and fortieth, perhaps not the four thousand eight hundredth part of what they are at present capable of performing, in consequence of a proper division and combination of their different operations.'
Smith favoured division of labour. He suggests that there are three causes of increase in the quantity of work:

- Increase in dexterity in every particular workman: The division of labour reduces every man's business to some one simple operation, and by making this operation the sole employment of his life, necessarily increases very much the dexterity of the workman.
- Saving the time which is commonly lost in passing from one species of work to another: He suggests that it is impossible to pass very quickly from one kind of work to another that is carried on in a different place, and with quite different tools. A country weaver, who cultivates a small farm, must lose a good deal of time in passing from his loom to his field, and from the field to his loom. When the two trades can be carried on in the same workhouse, the loss of time is no doubt much less.
- Invention of a great number of machines which facilitate and abridge labour, and enable one man to do the work of many: According to Smith, a greater part of the machines made use of in manufactures in which labour is most subdivided, were originally the inventions of common workmen, who, being each of them employed in some very simple operation,

naturally turned their thoughts towards finding out easier and readier methods of performing it.

1.4 Theory of Value

A main concern for Smith involved tracing the roots of value. He identified two different kinds of value, 'use value' and 'exchange value.' The concept of exchange value interested Smith considerably. The diamond-water paradox, in particular, proved puzzling to him: Why is it that diamonds, which have very little practical use, command a higher price than water which is indispensable to life? By discovering the true source of value Smith hoped to find a benchmark for measuring economic growth. Eventually Smith settled on labour as the source of value: The number of hours labour that a good can be exchanged for constitutes its inherent worth. (Note, this is not the same as saying that a good is worth the number of hours spent in its production.) The value of a good can also be referred to as the 'natural price.' The natural price need not function as the actual cost of a good in the marketplace. Competition, however, was expected to push the market price towards the natural price. Smith believed that the word value has two different meanings, and sometimes expresses the utility of some particular object, and sometimes the power of purchasing other goods which the possession of that object conveys. The one may be called 'value in use'; the other, 'value in exchange.' The things which have the greatest value in use have frequently little or no value in exchange; and on the contrary, those which have the greatest value in exchange have frequently little or no value in use.Nothing is more useful than water: but it will purchase scarce anything; scarce anything can be had in exchange for it. A diamond, on the contrary, has scarce any value in use; but a very great quantity of other goods may frequently be had in exchange for it.

According to Smith, value in exchange is the power of a commodity to purchase other goods –its price. This is an objective measure expressed in the market. His concept of value in use is ambiguous; it resulted in a good part of his difficulties in explaining relative prices. On the one hand, it has ethical connotations and is therefore a return to scholasticism. Smith's own puritanical standards are particularly noticeable in his statement that diamonds have hardly any value in use. On the other hand, value in use is the want-satisfying power of a commodity, the utility received by holding or consuming a good. Several kinds of utility are received when a commodity is consumed: its total utility, its average utility, and its marginal utility. Smith's focus was on total utility -the relationship between marginal utility and value was not understood by economists until one hundred years after Smith wrote - and this obscured his understanding of how demand plays its role in price determination. It is clear that the total utility of water is greater than that of diamonds; this is what Smith was referring to when he pointed to the high use value of water as compared to the use value of diamonds. However, because a commodity's marginal utility often decreases as more of it is consumed; it is quite possible that another unit of water would give less marginal utility than another unit of diamonds. The price we are willing to pay for a commodity -the value we place on acquiring another unit - depends not on its total utility but on its marginal utility. Because Smith did not recognize this (nor did other economists until the 1870s), he could neither find a satisfactory solution to the diamond-water paradox nor see the relationship between use value and exchange value. Smith's 'labour theory of value' stipulates that the value of a good or service is dependent upon the labour used in its production. It suggests that goods which take the same amount of time to produce should cost the same. This theory was an important concept in the philosophical ideals of Karl Marx. Opponents of the labour theory of value purport that it is not labour that determines the price of a good or service; rather, it is simply a function of supply and demand for a given good or service that determines its price. According to the theory, if the cost of purchasing something is greater than the amount that the purchaser values the time it would take to produce the good, then he will make it himself rather than buy it.

1.5 Laissez-Faire

Laissez-faire describes an economic system in which there is minimal intervention from governments and regulations. (The term means 'leave it alone' in French.) Advocates of laissez-faire assert that transactions between private parties should be free from regulations, taxes, tariffs, and monopolies, and that the role of government should be limited to policing against fraud and enforcing contracts. Adam Smith, who argued that individuals who pursue their own desires are able to best contribute to society as a whole, and this view was widely accepted in the 19th century. Adam Smith coined the term 'Laissez-Faire' in his book 'The Wealth of Nations' to describe an economic system in which the markets are completely unregulated and individuals are economically free to do as they please with no government interference. Historically, laissez-faire was a reaction against mercantilism, a system of commercial controls in which industry and trade, especially foreign trade, were merely seen as means of strengthening the state. Navigation laws, trade monopolies, taxes, and paternalistic regulations of all kinds bore heavily upon the rising class of merchants in the period of European colonial expansion. It was on behalf of this class that the French physiocrats, pioneer economists in the 18th cent., first formulated the principles of laissez-faire. With the physiocrats, state non-interference became a cardinal teaching; they especially opposed the taxation of commercial pursuits. Thus Laissez-faire economics originated in the 18th century France where economists, then called 'physiocrats', such as Francois Quesnay and Mirabeau became hostile to subsidies and discriminatory economic measures of then prevalent mercantilism.

Opposition to mercantilism and state paternalism also motivated Adam Smith, father of classical economics, whose name more than any other is connected with British laissez-faire doctrines. Smith believed that individual welfare rather than national power was the correct goal; he thus advocated that trade should be free of government restrictions. When individuals were free to pursue self-interest, the 'invisible hand' of rivalry or competition would become more effective than the state as a regulator of economic life. Smith did not believe in laissez-faire in an absolute sense; he found a place for government activity in public works, such as the building of canals and docks to facilitate trade, and in the regulation of foreign commerce to protect certain home industries.

Laissez-faire waned in popularity in the 20th century, when it proved inadequate to deal with social problems caused by industrialization. In 19th century, this philosophy became the dominant economic thinking in the Western world. But soon its failings reflected in the great disparity in distribution of wealth, harsh treatment of workers, disregard for consumer safety, and spread of monopolies, became clear. By the mid 19th century, opposition to laissez-faire economics began and governments in all industrialized countries intervened on behalf of workers and general population.Factory laws and consumer protection laws were enacted and growth of monopolies was checked. Early 20th century saw breakup of monopolies in the US and (after Second World War) nationalization of essential industries and services in Europe. Keynesian economics (which advocates government intervention in the national economy) further undermined it, a process spurred by the great depression of 1930s. From 1970's, however, the pendulum swung back to laissez-faire economics (renamed 'market economy' or 'free enterprise' in the disguise of globalization) and brought deregulation of business, and progressive removal of trade barriers, which is continuing.

1.6. David Ricardo's Works

Ricardo read Adam Smith's Wealth of Nations (1776) when he was in his late twenties. This sparked an interest in economics that lasted his whole life. In 1809 Ricardo began to write down his own ideas in economics for newspaper articles. In his Essay on the Influence of a Low Price of Corn on the Profits of Stock (1815), Ricardo articulated what came to be known as the law of diminishing returns. (This principle was also discovered simultaneously and independently by Malthus, Robert Torrens and Edward West). In 1817 David Ricardo published Principles of Political Economy and Taxation. In this text Ricardo integrated a theory of value into his theory of distribution. David Ricardo's attempts to answer important economic issues took economics to an unprecedented degree of theoretical sophistication. He outlined the Classical system more clearly and consistently than anyone before had done. His ideas became known as the 'Classical' or 'Ricardian' School. While his ideas were followed they slowly were replaced. However, even today the 'Neo-Ricardian' research program exists.

Modern discussion of the validity of the economic system of David Ricardo has centred about the theory of value. The labour theory of value states that the relative price of two goods is determined by the ratio of the quantities of labour required in their production. His labour theory of value has then following assumptions:

- ☆ Both sectors have the same wage rate and the same profit rate;
- ☆ The capital employed in production is made up of wages only;
- ☆ The period of production has the same length for both goods.

Ricardo himself realized that the second and third assumptions were quite unrealistic and hence admitted two exceptions to his labour theory of value:

I. Production periods may differ;

II. The two production processes may employ instruments and equipment as capital and not just wages, and in very different proportions.

Ricardo continued to work on his value theory to the end of his life. The first chapter serves merely as an introduction to a long book that debates an extended series of comparisons and which contrasts the various points of view while outlining Ricardo's own reasoning. In the chapter 'On Value and Riches,' Ricardo makes effort to illustrate that exchange value is not the same as 'value in use'. In this way one can factor two often contradictory results. Point 2, above, that the capital employed in production must be made up of wages only for his value theory to hold, is answered by this: that production may be made up of capital and machinery, but it doesn't change the principle (which he attributes to Adam Smith) that he tries to lay out in this chapter. Machinery may add to one measure of value beyond almost all measure without adding one penny to the other measure of value. In this way, one is able; Ricardo seems to show, to factor out somewhat contradictory assumptions which if confounded lead to equally contradictory results. By making all things perfectly clear, or in attempting to, Ricardo, seeks to resolve some of those ills of the democratic society in which he lived in so far as reason, and action, could resolve them. In this pursuit he gave stirring and amusing speeches to Parliament in order to change inner policies of the British Empire. The key point that Ricardo seems to make is: Accumulation of capital adds riches without decreasing the value of things to be traded, which may bring the various economic factors to a win-win. Ricardo first attempts to show that new riches are not adding as much value as one would think because they are always decreasing somewhat from the exchangeable value of what was produced. The decreasing value in exchange as value-in-use increases he extrapolates to infer that the sum world total of value in exchange is a fixed constant. Therefore, in the growth of the global economy, the first-world countries, he states, will begin to lose value per trade, even to the purely theoretical extent of taking from the capital base. Yet, Ricardo notes that with more value-in-use for the rich and the poor, both will likely obtain more security as the aggression of competition is mitigated by physical economic growth.

Adam Smith had thought that due to its effect on value, the growth of wealth of the poor beyond subsistence levels is likely to take from the overall wealth of the society. All economists (neoliberal through progressive) still worry about that and so they weaken the wealth of the poor to maintain economic growth. Ricardo shows this as unnecessary when we measure value in exchange together with the growth of value-in-riches, rather than by the monopolization value. The extreme aspects of competition then leave, for the rich and poor alike, an appearance of the growth of wealth, yet without the actual result of it. Taking a step back and noticing the growth of actual value-in-use may allow people as corporations and labourers, both rich and poor, to realize this and see a way and means forward.

David Ricardo maintained that the economy generally moves towards a standstill. His analysis is rooted in a modified version of the labour theory of value. He held out the belief that the rate of profit for society as a whole depends on the amount of labour necessary to support the workers who farm 'the most barren land that can still maintain agriculture' This model breaks land down into categories based on average fertility rates. The most fertile land naturally produces more food than land of poorer quality. As a result it commands a higher rent. The poorest land utilized for agriculture receives no rent, with all of its earnings going to cover labour and capital costs. The difference between the output from the least fertile land which can still be farmed and that of a higher quality constitutes the source of rent on the better land. As the population grows, poorer land must be cultivated in order to meet the growing demand. The cost of rent for good land then increases.

This coupled with the fact that poor land necessitates increased labour input to maintain minimal output results in falling profit levels. As rents rise, profits fall. Essentially, rent costs gobble up profits as the population increases. Since profits lead to reinvestment and thus growth rising rent costs indirectly prevent economic progress. In Ricardo's model the interests of landowners directly oppose those of general society. Ricardo pre-empted Karl Marx in describing adversarial class relations. Marx, in fact, based a great deal of his economic theory on Ricardo's writings. Although Marx identified capitalists, not landlords, as the source of societal grief he co-opted Ricardo's labour theory of value. Marx also utilized Ricardo's forecast of economic stagnation in predicting a working class uprising. When capitalism eroded its own underpinnings the resulting misery was expected to bring social strife and revolution. It is unlikely that Ricardo would have supported Marx's revolutionary brand of political economics, but the ties between the schools of thought are undeniable. In the Ricardian system the theory of value, reduced to Ricardo's level of simplification, plus the theory of rent, provided the key to the central problem of income distribution. It was, of course, necessary to relate the theory of value to the theory of prices in a complex economy. Ricardo did this by relating market price to the costs of production in the marginal (no-rent) firm. He noted: The exchangeable value of all commodities, whether they be manufactured, or the produce of the mines, or the produce of the land, is always regulated, not by the less quantity of labour that will suffice for their production under circumstances highly favourable, and exclusively enjoyed by those who have peculiar facilities of production; but by the greater quantity of labour necessarily bestowed on their production by those who continue to produce them under the most unfavorable circumstances. Ricardo recognized that there is no perfect measure of value, since any measure chosen varies with fluctuations in wages and profit rates. We have seen that different durability of capital and different ratios of fixed to circulating capital will affect market prices differently if wages change relative to profits. Thus Ricardo devised an analytical gimmick -the 'average firm'- in which both the ratio of capital to labour and capital durability are assumed equal to the economy average. So armed, Ricardo was ready to solve the problem of income distribution and its changes over time.

Mr. Malthus shows that in fact the exchangeable value of commodities is not exactly proportioned to the labour which has been employed on them, which I not only admit now, but have never denied. Did Ricardo have a labour theory of value–did he believe that the relative values of commodities are governed exclusively by the relative quantities of labour necessary to produce them? A considerable number of historians of economics have given a fiat affirmative answer to this question–a surprisingly large number considering the fact that there is not the slightest basis for such an answer.

1. In the course of their expositions one encounters quite remarkable statements such as that Ricardo assumed that labour and capital were in a fixed proportion in all industries,
2. or that "Ricardo. constantly takes no notice of capital."
3. Presumably these writers did not have access to Ricardo's Principles.

More careful historians of doctrine have recognized the several and important departures from a pure labour theory that Ricardo emphatically made. There is, in fact, an almost continuous spectrum of interpretations. At one uninteresting extreme some writers argue that Ricardo simply forgot or did not understand the import of the qualifications he made. A very important group has advanced the view that Ricardo wished to hold a labour theory of value–Cannan heads his treatment:"Ricardo's Attempt to Revive the Pure Labour Theory." They hold that under adverse criticism and honest self-examination, Ricardo was gradually forced to introduce in successive editions of the Principles a series of qualifications of increasing importance, so that in the end it really was not a labour theory.But Ricardo was not willing to abandon the theory completely: "his heart clung to the pure labour theory,"6 or he made a "brave show" of "identifying, as far as might be, value and the amount of labour necessary for production."In the most sophisticated versions of this view, the formal exposition of Ricardo's theory is, as I believe, correct and complete in substance, but there persists a strong implication that Ricardo attributes more than quantitative importance to labour in determining values.8 Thus, after carefully stating Ricardo's "modifications" of a labour theory, St. Clair says that Ricardo "sweeps them all into the waste-paper basket"; for "he never got rid entirely of the idea with which he started, namely, that labour is the only price exacted by Nature for her gifts."The only economists to argue at some length that Ricardo had a cost-of-production theory of value, so far as I know, have been Marshall, Diehi, and Viner–but who could wish for more comforting allies?'The widespread misinterpretation of a leading doctrine of an economist of the first rank is not a product only of later viewpoints and knowledge, for it occurred already in Ricardo's lifetime. The present essay seeks to set forth precisely what Ricardo's theory of value was, and to examine the interpretation placed upon it by his leading contemporaries.Ricardo's formulation of his theory of value was much influenced by his desire to correct what he believed to be the major errors of Adam Smith's theory. For Smith the long-run value of a commodity equaled its cost of production: the "natural" price of a commodity was the sum of the necessary payments for labour, capital, and land. A rise in the price of one of these factors, and in particular a rise in wages, would lead to a rise in the prices of the commodities in which the factor entered." If the changes in values were to be more than nominal price- level changes," however, this was clearly a superficial analysis: why should the relative values of commodities be affected in any systematic way by the absolute level of input prices? The organization, although not the content, of Ricardo's chapter on value can be interpreted as one which presses the criticism of Smith's theory to the utmost limits.

The analysis is limited to useful goods, produced in free competition, and the element of rent is temporarily put aside (and later shown not to enter into marginal cost). Ricardo begins with the simplest case: the commodities are produced by one type of labour alone, working perhaps on free land.In this simplest case the relative values of commodities will clearly equal the relative quantities of labour necessary to produce them, and will be wholly unaffected by the absolute level of wages (no matter in what unit they are measured).Consider next, with Ricardo, the case in which only labour is required to produce the commodities, but different types of labour are used in differing proportions.The market will establish wage differentials corresponding to the differences in skill and training of the occupations, and "the scale, when once formed, is liable to little variation." Hence a rise of wages will affect the money costs of all commodities in equal proportion, and leave relative values unaffected. Ricardo did not consider the possibility that the relative amounts of skilled and unskilled labour employed to produce a commodity might change and hence its relative value would change; he could have asserted, however, that the relative value of the commodity will change only if the "common labour" equivalent of the original labour input changed."

Ricardo quoted as a striking example of this argument the passage: "By regulating the money price of all the other parts of the rude produce of land, [the price of corn] regulates that of the materials of almost all manufactures. By regulating the money price of labour, it regulates that of manufacturing art and industry. And by regulating both, it regulates that of the complete manufacture. The money price of labour, and of every thing that is the produce either of land or labour, must necessarily either rise or fall in proportion to the money price of corn"."And Ricardo was not inclined to make this exception since he had a commodity theory of money."But a closer analysis would have indicated that the wages of superior labour contain interest on the investment in acquiring skill, and therefore the relative levels of wages and interest rates enter into relative values.It s evident that we can still retain the proposition that relative values of commodities are independent of the absolute level of wages (and profits) if, when each worker is equipped with fixed capital, we assume that the ratio of fixed capital to labour is the same in every industry, provided the capitals have equal durability. And this is Ricardo's next case (I, 26 if.). At this point it is not made clear whether the fixed capital earns a net return: Ricardo views the fixed capital as previously expended labour, and says, correctly but unnecessarily, that "exchangeable value of the commodities produced would be in proportion to the labour bestowed on their production; not on their immediate production only, but on all those implements or machines required to give effect to the particular labour to which they were applied." At a later point it is made utterly clear that the contribution of the fixed capital consists of not only amortization quotas but also interest on the investment.'and this is as far as Ricardo could go in attacking Smith's theory. The next step, and analytically the final step, is to allow the ratio of fixed capital to labour to vary between commodities, and when this is done: This difference in the degree of durability of fixed capital, and this variety in the proportions in which the two sorts of capital [fixed and circulating] may be combined, introduce another cause, besides the greater or less quantity of labour necessary to produce commodities, for the variations in their relative values– this cause is the rise or fall in the value of labour.

Ricardo believed that the changes brought about in the relative values of commodities by fluctuations in wages and profits were very small relative to those brought about by fluctuations in the quantity of labour (direct and indirect): The reader, however, should remark, that this cause of the variation of commodities is comparatively slight in its effects. With such a rise of wages as should occasion a fall of one per cent. in profits, goods produced under the circumstances I have supposed, vary in relative value only one per cent.; they fall with so great a fall of profits from 6,050 1. to 5,995 1. The greatest effects which could be produced on the relative prices of these goods from a rise of wages, could not exceed 6 or 7 per cent.; for profits could not, probably, under any circumstances, admit of a greater general and permanent depression than to that amount?And so, although it would be "wrong wholly to omit the consideration of the effect produced by a rise or fall of labour [wages], it would be equally incorrect to attach much importance to it" and therefore in the remainder of the book he will "consider all the great variations which take place in the relative value of commodities to be produced by the greater or less quantity of labour which may be required from time to time to produce them".

I can find no basis for the belief that Ricardo had an analytical labour-theory of value, for quantities of labour are not the only determinants of relative values. Such a theory would have to reduce all obstacles to production to expenditures of labour or assert the irrelevance or nonexistence of noulabour obstacles, and Ricardo does not embrace either view. On the other hand, there is no doubt that he held what may be called an empirical labour theory of value, that is, a theory that the relative quantities of labour required in production are the dominant determinants of relative values. Such an empirical proposition cannot be interpreted as an analytical theory, any more than the now popular view that the price level is governed by the wage level and the productivity of labour can possibly be defended as an analytical proposition.This is not to say that Ricardo's analytical theory was correct, for it contained several important deficiencies. It excluded rent from costs, and even if the supply of land were fixed the rent a piece of land could yield in one use would be a cost to other uses. (Ricardo's practice of assuming that land was used to grow only corn obscured this point.) His theory was wrong in reducing all capital to previously expended labour plus interest; except in some irrelevant day of Genesis all capital has been made by the cooperation of earlier capital and labour and land. This view may have fostered his empirical judgment that labour.

Ricardo's Principles received very diverse reviews, ranging from the adulation of McCulloch to the reaction of one anonymous reviewer that the volume "contains no valuable information in point of fact, and very little good reasoning in point of doctrine." This same diversity extended to the interpretation of his theory of value. B. Say could find only a simple labour theory of value. In the notes he added to the French translation of the Principles, he observed. Ricardo does not appear to have included [in the contribution of machinery to the value of a commodity the profits or the interest on the capital as constituent parts of the prices of commodities.If to this cause of variation [differences in rate of durability of capitals] we add the exception noticed by Mr. Ricardo, arising from the greater or less proportion of fixed capital employed in different commodities, the effects of which would shew themselves in a very early period of savage life; it must be allowed that the rule which declares "that commodities never vary in value unless a greater or less quantity of labour be bestowed on their production," cannot possibly, as stated by Mr. Ricardo, be "of universal application in the early stages of society."It should be noticed that apropos of this discussion, Ricardo says, "In all the observations of Mr. Maithus on this subject I most fully concur". The only difference between Mathus and Ricardo in their concepts of costs of production was that the former included and the latter excluded the rent of land. The real dispute between themWhen James Mill wrote his primer on the Ricardian economics, Elements of Political Economy (1821), he restated Ricardo's theory in substance. After stating a labour-quantity theory, he went on to explain at length that because of differences in capital-labour ratios in various industries the fluctuations in wages and profits affect exchange values. He concluded:It is evident, however, that though this difference in the ratios according to which the wages of two kinds of labour were exchanged, and the different proportions in which they were applied in the production of commodities, would, upon a rise or fall in wages, alter the relative value of commodities, it would do so, without in the least affecting the truth of the previous proposition, that quantity of labour determined exchangeable value.

Mr. Ricardo, indeed, explicitly allows the influence of other causes, such as time, differences in the proportion of fixed and circulating capital, and inequalities in the durability of capital, by which he admits the value of commodities is liable to be affected, Notwithstanding these modifications, however, his followers continue to lay down the position of quantity of labour being the sole cause of value in the most precise and positive terms; not that they deny the exceptions, but they appear to lose sight of their existence, and frequently fall into language incompatible with teir admission; It may be added that in substance Bailey accepts Ricardo's theory of value, including the exclusion of rent from costs of production.

We may recapitulate this brief survey. McCulloch, Bailey and Maithus correctly understood Ricardo's theory to be a cost-of -production theory excluding rent, and De Quincy should probably be added to this group. The theory was understood as a simple labour-quantity theory by Say and Mill, and also by Torrens." It is worth repeating that Ricardo accepted Malthus' analysis and rejected Mill's. The theory was more widely understood in its correct sense in Ricardo's time than in later times.

How did the misunderstanding of Ricardo's theory arise? Although Ricardo's exposition has been often and justly denounced, the main argument stands out clearly enough: it does not require great generosu A Critical Dissertation on tue Nature, Measures, and Causes of Value, London 1825. "Mr. Ricardo has pushed this principle still further, and contended, that in all period of society, whether before or after the accumulation of capital and appropriations of land, the labour expended upon production is the sole regulator of value." In Ricardo's period several factors probably played a minor role in the confusion. Ricardo's two leading disciples, Mill and McCulloch, asserted a labour-quantity theory with all emphasis, although actually neither held such a theory to the extent of denying that fluctuations in wage and profit rates affected commodity values. Their expositions naturally colored the interpretation of Ricardo, even though McCulloch expressly indicated his disagreement with Ricardo. Another source was the vast confusion of the causes of value with the proper measure of value, and in Ricardo's first edition a pure labour measure of value was used.

The main source of the confusion, however, was probably the failure of economists to distinguish clearly between analytical and empirical propositions. Among economists who were not methodologically self- conscious, who did not systematically consider the necessary and sufficient conditions for equilibrium, the distinction would seldom be remarked. Ricardo's emphasis upon the quantitative importance of labour tended to be read as an analytical proposition that labour quantities were the sole regulators of value.The failure to distinguish between analytical and empirical propositions has been a source of much misunderstanding in economics. An analytical statement concerns functional relationships; an empirical statement takes account of the quantitative significance of the relationships. When Marshall viewed the demand for a commodity as a function of its price, the prices of closely related goods, and of income, he was criticized by members of the Wairasian school for failing to recognize that all prizes in principle influence the demand for any commodity. This is a characteristic instance of the distinction in question:No Marshallian ever denied the existence of the formal relationships that were omitted; no Wairasian ever presented an empirical example of important error resulting from their neglect.

Schumpeter asked why if Marshall's (and the present) interpretation of Ricardo is correct, there should have been any controversy–would it then not amount simply to the current cost-of-production theory? One is inclined to reply that there was no controversy, and that the controversy was about something else. There was in fact no active controversy over the so-called labour theory in Ricardo's lifetime. The main points of controversy were different. First, Ricardo eliminated rent from the costs of production, which was not in keeping with popular views. Second, he appeared to deny (but did not do so) that supply and demand governed value; in fact he considered this a wholly superficial view that merely postponed analysis of the real determinants of relative values, namely the factors governing supply. Finally, the endless dispute between Maithus and the Ricardians concerned the measure of value, not its causes.

The basic reason Ricardo's theory is often misinterpreted is that it was often misinterpreted in the past. If a theory once acquires an established meaning, each generation of economists bequeaths this meaning to the next, and it is almost impossible for a famous theory to get a fresh hearing. Perhaps one hearing is all that a theory is entitled 'to, but one may plead that Ricardo deserves at least a rehearing–his theory is relatively more widely misunderstood today than it was in his lifetime. One can build a strong case that the modern economist need not be acquainted with Ricardo's work, but there is no case for his being acquainted with an imposter.

4

The Classical Labour Theory of Value

4.1 The Classical Labour Theory of Value

Either the labour theory of value, or, secondarily, some other form of cost theory of value, was common to the classical school of political economy in England. It was stated by Adam Smith in ambiguous form: 'The real price of everything, what everything really costs to the man who wants to acquire it, is the toil and trouble of acquiring it. Labour was the first price, the original purchase-money that was paid for all things.' In the same passage, though, he spoke of the value of a commodity in one's possession as consisting of 'the quantity of the labour which he can command.' And at other times, he seemed to make the market price of labour the source of its effect on exchange value.

The most clear-cut and effective statement of the labour theory was by David Ricardo, in Principles of Political Economy and Taxation: 'The value of a commodity, or the quantity of any other commodity for which it will exchange, depends on the relative quantity of labour which is necessary for its production, and not as the greater or less compensation which is paid for that labour.' In so defining the doctrine, Ricardo eliminated the confusion between labour as the source of exchange-value and wages as a component of price.

From this principle, it followed that income accruing to the owners of land and capital was a deduction from this exchange-value created by labour, and that wages varied inversely with profit: 'If the corn is to be divided between the farmer and the labourer, the larger the proportion that is given to the latter, the less will remain for the former. So if cloth or cotton goods be divided between the workman and his employer, the larger the proportion given to the former, the less remains for the latter.'

It was only natural that the emerging socialist movement should seize on the political implications of this conclusion. The school of so-called 'Ricardian socialists' in England took just such an inspiration. The greatest of them, Thomas Hodgskin, wrote in 'Labour defended against the claims of capital', 'Wages vary inversely as profits, or wages rise when profits fall, and profits rise when wages fall; and it is therefore profits, or the capitalist's share of the national produce, which is opposed to wages, or the share of the labourer.'

Marx, in turn, was inspired by the Ricardian socialist interpretation of classical political economy, as well as by Proudhon. According to Engels, modern socialism was a direct outgrowth of the insights of 'bourgeois political economy' on the nature of wages, rent, and profit. Insofar as modern socialism, no matter of what tendency, starts out from bourgeois political economy, it almost without exception takes up the Ricardian theory of value. The two propositions which Ricardo proclaimed in 1817 right at the beginning of his Principles, 1) that the value of any commodity is purely and solely determined by the quantity of labour required for its production, and 2) that the product of the entire social labour is divided among the three classes: landowners (rent), capitalists (profit), and workers (wages)–these two propositions had ever since 1821 been utilized in England for socialist conclusions, and in part with such pointedness and resolution that this literature, which had then almost been forgotten and was to a large extent only rediscovered by Marx, remained surpassed until the appearance of Capital.The actual extent to which Marx's theory of value is a straightforward outgrowth of Ricardo's, and to which it was a pre-existing Hegelian philosophy with Ricardian elements grafted on, is an issue in dispute. But for the present purpose, we will treat Marx's theory of value as relevant to our study to the extent that it is amenable to a Ricardian approach.

Jean-Baptiste Say (5 January 1767 – 15 November 1832) was a French economist and businessman. He had classically liberal views and argued in favour of competition, free trade, and lifting restraints on business. He is best known due to Say's Law, which is named after him and at times credited to him, but while he discussed and popularized it, he did not originate it. J. B. Say was born in Lyon. His father, Jean-Etienne Etienne Say, was of Protestant family which had moved from Nîmes to Geneva for some time in consequence of the revocation of the Edict of Nantes. (His brother Louis Auguste (1774–1840) was also an economist). Say was intended to follow a commercial career, and was sent, with his brother Horace, to England: here he lived first in Croydon, in the house of a merchant, to whom he acted as clerk, and afterwards in London, where he was in the service of another employer. When, on the death of the latter, he returned to France, he was employed in the office of a life assurance company directed by Étienne Clavière. Say's first literary attempt was a pamphlet on the liberty of the press, published in 1789. He later worked under Mirabeau on the Courrier de Provence. In 1792 he took part as a volunteer in the campaign of Champagne; in 1793 he assumed, in conformity with the Revolutionary fashion, the pre-name of Atticus, and became secretary to Clavière, then finance minister. In 1793 Say married Mlle Deloche, daughter of a former lawyer. From 1794 to 1800 Say edited a periodical entitled La Decade philosophique, litteraire, etpolitique, in which he expounded the doctrines of Adam Smith. He had by this time established his reputation as a publicist, and, when the consular government was established in 1799, he was selected as one of the hundred members of the tribunate, resigning the direction of the Decade. In 1800 he published in Olbie, ouessaisur les moyens de reformer les moeursd'une nation. In 1803 appeared Say's principal work, the Traitéd'économiepolitiqueou simple exposition de la manièredont se forment, se distribuentet se composent les richesses. In 1804, having shown his unwillingness to sacrifice his convictions for the purpose of furthering the designs of Napoleon, he was removed from the office of tribune. He then turned to industrial pursuits, and, having made himself acquainted with the processes of the cotton manufacture, founded at Auchy, in the Pas de Calais, a spinning-mill which employed four or five hundred persons, principally women and children. He devoted his leisure to the improvement of his economic treatise, which had for some time been out of print, but which the censorship did not permit him to republish.

4.1.1. Say's Methodology

Say's approach to economics is, in philosophical terms, that of a realist and an essentialist. He combines a healthy scepticism regarding the usefulness of statistical investigations with an emphasis on observing the facts of reality. A statistical description 'does not indicate the origin and consequences of the facts it has collected.' For Say, only a causal analysis based on the essential natures of the entities involved can achieve that end, and such an analysis is the core task of political economy. He sees economics as a genuine science capable of establishing 'absolute truths,' but insists that it 'has only become a science since it has been confined to the results of inductive investigation.' In fact, Say declares that political economy 'forms a part of experimental science' and is, thus, rather similar to chemistry and natural philosophy. Taxonomically, he divides all facts into (a) those that refer to objects and (b) those that refer to events or interactions. The former is the domain of descriptive science (*e.g.*, botany); while the latter is the domain of experimental science (*e.g.*, chemistry or physics).Above all, Say seeks to be practical; for 'nothing can be more idle than the opposition of theory to practice!' To that end, he attempts always to employ language that is precise and yet as simple as possible, so that any literate, reasonably intelligent person can comprehend his meaning. For Say, as for most modern Austrians, economics is not a shadowy realm to be penetrated only by the expert, but a subject of enormous practical importance accessible to all. It is thus no surprise to find that Say, in keeping with such a goal of lucidity and intelligibility, criticizes Adam Smith's Wealth of Nations for being 'destitute of method,' obscure, vague, and disjointed as well as for containing too many long and distracting digressions on topics such as war, education, history, and politics.

4.1.2. Say's Law of Market

Say's law is commonly summarized as 'supply creates its own demand.' This law, also referred to as Say's 'theory of markets' or 'law of markets,' indicates that the act of producing aggregate output generates a sufficient amount of aggregate income to purchase all of the output produced. This principle indicated that excess production or insufficient demand for production was unlikely to occur, at least for any extended period. When combined with flexible prices and saving investment equality, Say's law further implied that an economy would achieve and maintain full employment of resources. This law was singled out by John Maynard Keynes in his critique of classical economics, but remains relevant in current macroeconomic analysis, reflected in the circular flow model. Say's law that 'supply creates its own demand' is one of the fundamental principles of classical economics. This principle though is attributed J B Say, was simultaneously advocated by several economists of the day. The law is often misinterpreted to mean any firm that produces a good is guaranteed profitable sales. It more correctly means that the act of production generates enough income for an equivalent amount of demand for other production. When applied to the macro economy, Say's law indicates that economic downturns cannot be caused by the lack of demand, that is, overproduction. The aggregate production of output generates just enough income to equate aggregate demand with aggregate supply.

This principle was, and still is, one of the key building blocks of the classical school of economics. It remains, in some guise or other, essential to any defence of free markets. Moreover, all collectivists attempt to refute it in the course of their assault on liberty and the free society. And yet, some writers have questioned the profundity of Say's Law. Alexander Gray refers to 'this theory, which perhaps does not come to much.' Say identifies two means by which the corrective process operates. Principally, he argues that, though individuals do save part of the income derived from production, as long as those savings are reinvested in 'productive employment,' in the aggregate there need be no decreases in production, income, or consumption. This process of reinvestment is fuelled by differences in the profits earned by entrepreneurs. Those goods that are relatively more scarce, and thus rising in price, attract additional investment, while those that are relatively less scarce, and thus falling in price, discourage investment. And even if one hoards money or buries it, 'the ultimate object is always to employ it in a purchase of some kind,' so there still cannot be deficient demand as long as real economic values are being produced. In order for consumers to exist, there must first be producers. Throughout his discussion of production and consumption, Say consistently maintains that money is merely a neutral conduit through which aggregate supply is translated into aggregate demand, or 'money is but the agent of the transfer of values.' There seems to be no recognition of the transmission mechanism by which changes in the supply of money alter the relative prices of goods and, thereby, redirect the entire interrelated structure of production. From a modern Austrian perspective, Say's failure to grasp the non-neutrality of money must be deemed a deficiency of some note.

On the other hand, Say eloquently expresses a clear understanding that it is wholly beneficial for a society to experience generally falling prices whenever such declining prices are the result of productivity gains. Not only does this circumstance indicate, contrary to popular belief, 'that a country is rich and plentiful,' but also that 'products formerly within reach of the rich alone have been made accessible to almost every class of society.' Moreover, Say correctly perceives that (a) the prices of goods reflect their utility to the buyer, (b) the prices of the factors of production are derived or 'imputed' from the prices of the goods produced, and therefore (c) costs of production represent an interface between the utility of the good and the productivity of the factors of production. Please note this: Most textbooks abbreviate Say's Law into the transparently false proposition 'supply creates its own demand.' At minimum, this should be given as 'aggregate supply creates its own aggregate demand,' because the claim is not that the production of commodity X necessarily results in an equivalent demand for X, but that the production of X leads to demand for commodities A, B, C, and so forth. The production, or supply, of commodities (and complementary services) in general leads to the consumption of, or demand for, commodities (and complementary services) in general. It is certainly possible for there to exist either a shortage or a surplus of any particular commodity, but general overproduction or general underproduction can be no more than momentary phenomena. 'It is because the production of some commodities has declined, that all other commodities are superabundant,' and such maladjusted production results from 'some violent means a political or natural convulsion.' Left to its own devices, the market will correct such imbalances.

4.2. Malthus (Thomas Robert Malthus)

Thomas Robert Malthus was born near Guildford, Surrey in February 1766. His father was prosperous but unconventional and educated his son at home. Malthus went on to Cambridge University, earning a master's degree in 1791. In 1793, he was made a fellow of Jesus College, Cambridge. In 1805, Malthus became professor of history and political economy (the first holder of such an academic office) at the East India Company's college in Haileybury, Hertfordshire, where he remained until his death. In 1819, Malthus was elected a fellow of the Royal Society and two years later he became a member of the Political Economy Club, whose members included David Ricardo and James Mill. In 1824,he was elected as one of the 10 royal associates of the Royal Society of Literature. Malthus was also one of the co-founders of the Statistical Society of London in 1834. Malthus' most well-known work 'An Essay on the Principle of Population' was published in 1798, although he was the author of many pamphlets and other longer tracts including 'An Inquiry into the Nature and Progress of Rent' (1815) and 'Principles of Political Economy' (1820). The main tenets of his argument were radically opposed to current thinking at the time. He argued that increases in population would eventually diminish the ability of the world to feed itself and based this conclusion on the thesis that populations expand in such a way as to overtake the development of sufficient land for crops. Associated with Darwin, whose theory of natural selection was influenced by Malthus' analysis of population growth, Malthus was often misinterpreted, but his views became popular again in the 20thcentury with the advent of Keynesian economics. Malthus died on 23 December 1834.Malthus is best known for his hugely an influential theory on population growth. Malthus was the first economist to propose a systematic theory of population. He articulated his views regarding population in his famous book, Essay on the Principle of Population (1798), for which he collected empirical data to support his thesis. Malthus had the second edition of his book published in 1803, in which he modified some of his views from the first edition, but essentially his original thesis did not change.

4.2.1. Assumptions Thomas Robert Malthus

Malthus took into account two main assumptions:

1. Food is an essential component for human existence.
2. Humans have the basic urge to multiply.

Malthus' theory was based on the assumption that the power of population to multiply is much greater than the power of the earth to provide subsistence for man. In his own words 'passion between the sexes is an inevitable phenomenon', hence, when unchecked, population would grow at such a high rate that it would outstrip food supply. According to him, disease, food shortage and death due to starvation, were nature's way to control population. He proposed that human beings adopt measures like infanticide, abortion, delay in marriage and strict following of celibacy to check population growth. Malthus argued that because of the natural human urge to reproduce human population increases geometrically (1, 2, 4, 16, 32, 64, 128, 256, etc.). However, food supply, at most, can only increase arithmetically (1, 2, 3, 4, 5, 6, 7, 8, etc.). Therefore, since food is an essential component to human life, population growth in any area or on the planet, if unchecked, would lead to starvation. However, Malthus also argued that there are preventative checks and positive checks on population that slow its growth and keep the population from rising exponentially for too long, but still, poverty is inescapable and will continue. On the basis of a hypothetical world population of one billion in the early nineteenth century and an adequate means of subsistence at that time, Malthus suggested that there was a potential for a population increase to 256 billion within 200 years but that the means of subsistence were only capable of being increased enough for nine billion to be fed at the level prevailing at the beginning of the period. He therefore considered that the population increase should be kept down to the level at which it could be supported by the operation of various checks on population growth, which he categorized as 'preventive' and 'positive' checks.In his first edition of the essay, Malthus proposed two main solutions to the problem of population explosion, namely:

Positive Check

This method results in increase in death rate. He described this as God's way of restoring the Natural Order. Positive checks are those that increase the death rate. These include disease, war, disaster, and finally, when other checks don't reduce population, famine. Malthus felt that the fear of famine or the development of famine was also a major impetus to reduce the birth rate. He indicates that potential parents are less likely to have children when they know that their children are likely to starve. Although Positive Checks help in controlling the population growth, it brings with it widespread misery and pain. Hence, it is not regarded as an ideal solution to population problem.

Preventive or Negative Check

This method refers to human effort in reducing the birth rate. It is more practically and logically applicable. Abortion, prostitution, postponement of marriage, birth control and celibacy are few measures that were advised to be strictly followed in order to help solve the problem. In his second edition of the same essay, Malthus laid more emphasis on:

Moral Restraint

This is regarded as a universally applicable solution keeping up with the ideologies of virtue, economic gain and social improvement. According to this principle, one should refrain from marriage till the time he is capable of supporting a family with food, clothing and shelter. Until then he should follow strict celibacy. In the words of Geoffrey Gilbert, 'He (Malthus) went so far as to claim that moral restraint on a wide scale was the best means indeed, the only means of easing the poverty of the lower classes.'

Criticism

Some critics like Karl Marx, argued that Malthus failed to recognize the potential of human population to increase food supply. He is accused by many to have failed to comprehend man's ability to use science and technology to increase food supply to meet the needs of an increasing population. Thinkers from the field of social sciences have criticized Malthus for his belief that the human society could never be made perfect. He opposed all political, social and economic reforms that did not aim at controlling birth rate. His own methods of checking population growth were criticized for being impractical. Malthus was opposed to the Poor Laws popular in England, which provided relief to the people who qualified as poor under the laws. According to him, such charity would provide only short term relief to them and let the poor remain in their state of financial distress. This thought of Malthus was viewed as misanthropic. From his writings, some have interpreted Malthus as a rigid and pessimistic individual. However, he is also viewed by some as a pragmatic thinker, who put a check on the unbridled enthusiasm of some who viewed increase in population as a means of progress.

4.3. Theory of Glut

What Causes Gluts: In Britain in 1818, a large decline in agricultural prices led to the severe depression of 1819, and the high levels of unemployment that followed led to general social unrest: In August of that year, thousands of workers had demonstrated in the streets of Manchester. The British government had called out the armed forces and the demonstrators were brutally suppressed. Ten demonstrators were killed and many hundreds were severely injured in what came to be known as the 'Peterloo Massacre.' This occurred just one year before the publication of the first edition of Malthus's Principles. Malthus was extremely aware that depressions not only could but did happen in a capitalist economy; he was also well aware of the potential revolutionary danger of such labour uprisings. His single most important goal in writing the Principles was to promote an understanding of these crises, or gluts, and to propose policies to mitigate them. Malthus answer was to explain how it was possible for the demand for goods and services to be less than the amount of goods and services produced - so his was a theory of deficient demand – and when that happened, goods would pile up in the marketplace and a general glut would ensue. But why did this happen? Why did goods pile up in the marketplace and simply rot away (creatively) while so many people struggled to meet the necessities of life? His solution was to realize that the costs of production (*i.e.* the natural value of the goods) had to be equal to the income of the various classes in society, *i.e.* that the value of production had to equal wages plus interest plus profits (notice that rents are missing). He argued that labour is hired only if it produces a value greater than it is paid, so labour income alone won't be enough to consume all that is produced. That leaves Capitalists and landlords to make up the difference.

Capitalists' main desire was to accumulate capital, and to do that they needed to save. But the problem was that this led to a tendency for capitalists to save too much, more than there were profitable opportunities. Hence, some of the saving sat idle - imagine pieces of gold sitting in a safe somewhere doing nothing instead of being employed productively supporting new investments and this led to deficient (or, as he called it, ineffectual) demand. So capitalists did not consume enough. How to solve this? By moving income to landlords. Landlords could depend upon income from rent, and because of that, they consumed all that they earned – more rent would come in the next time period so they spent lavishly on 'personal services', there was no need to save. Importantly, this was unproductive consumption - landlords took from output but added noting to it - and hence they could be depended upon to make up the slack in demand if income could be shifted in their direction. There was no danger than this class would save too much. Here, then, was the solution. An important policy question of the day was whether the Corn Laws* - tariffs on cheaper grain from France and other countries - were helpful or not. Malthus answer was that they helped. By raising the price of corn (*i.e.* grains), rents would increase and this would shift income to landlords. They would then buy up all the extra goods the capitalists had left idle by saving too much and the problem would be solved. Britain and Ireland against competition from less expensive foreign imports between 1815 and 1846.Theeconomic issue, in essence, was food prices; the price of grain was central to the price of the most important food staple, bread, and the working man spent much of his wages on bread. The political issue was a dispute between and the new class of manufacturers and industrialists :the former desired to maximise their profits from agriculture, by keeping the price at which they could sell their grain high; the latter wished to maximise their profits from manufacture, by reducing the wages they paid to their factory workers–the difficulty being that men could not work in the factories if a factory wage was not enough to feed them and their families; hence, in practice, high grain prices kept factory wages high also.

4.4. J.S. Mill

John Stuart Mill was born in London on May 20, 1806 and died May 8, 1873 in Avignon, France.He was a philosopher, economist and political theorist. He was the eldest son of the historian and economist James Mill who, later assisted by Jeremy Bentham and David Ricardo, gave John Stuart his prodigious education. The rigorous upbringing did bear its fruit, by the age of 14 Mill was well versed in both Greek and Latin, and in the subjects of history, mathematics, logic and, what became his specialty, political economy. At the young age of 14, John Stuart Mill went to pursue his studies in France where he then met Henri Saint-Simon. Over the following few years he edited Bentham´s fragments on the theory of legal evidence. He then refused to study further in Oxbridge and was instead secured a position in the East India Company in 1823; he continued to work there until 1858. Mill had had a nervous breakdown at the age of twenty and suffered from severe depression. This is often considered a consequence of his rigorous education. He eventually recovered, in part due to his engagement with the poetry of William

Wordsworth. As well, during this time he distanced himself from Benthamism, working to add a more idealist note to it. He was also influenced by August Comte's work during this period and such reflections can be seen in Mill. Yet during this growth period Mill never entirely discarded utilitarian principles. Empiricist by nature, Mill instead worked to use such principles in a positive rather than a critical sense. This positivistic stance, he was convinced, was the way towards a real progress in human knowledge and the possibility of freedom to be attained.In 1865 Mill was elected as a Member of Parliament for Westminster, where he remained for three years. During this same period he also served as Rector of the renown University of St. Andrews. Mill was very pro-active and worked to implement reforms towards proportional representation and labour unions, and supported women's rights. He was the first parliamentary deputy to support a woman's right to vote and was a co-founder of the first women's suffrage organization. It was not until 1843 that John Stuart Mill became known as a philosopher. In this same year he published System of Logic, Ratiocinative and Inductive, his most systematic work.Whatever is known to us by consciousness, is known beyond possibility of question. What one sees or feels, whether bodily or mentally, one cannot but be sure that one sees or feels. No science is required for the purpose of establishing such truths; no rules of art can render our knowledge of them more certain than it is in itself. There is no logic for this portion of our knowledge. But we may fancy that we see or feel what we in reality infer.

Attacking 'intuitionist' philosophy, he argues in favour of logic as the most adequate method of proof. Despite the fact that truth 'may seem to be apprehended intuitively,' Mill stresses the fact that, 'it has long been ascertained that what is perceived by the eye, is at most nothing more than a variously coloured surface.' It thus the object of logic to 'distinguish between things proved and things not proved, between what is worthy and what is unworthy of belief.'

In 1848, Mill published Principles of Political Economy, which soon became the most important text of his time. The book examines the conditions of production, namely labour and nature. Following Ricardo and Malthus, he emphasizes the possibility of change and social improvement and examines environmental protection needs. In order for these to be obtained, he considers a limitation of both economic growth and population growth, as the polis itself is indispensable. Furthermore, Mill argued in favour of worker-owned cooperatives, which clearly reflect his views. On Liberty, published in 1859, caused the greatest controversy of John Stuart Mill's career and has since become a classic of liberal thought. Written and developed in close collaboration with his wife, Harriet Taylor, Mill examines the nature of power and argues for an absolute freedom of thought and speech. For Mill it is only through such 'freedom' that human progress can be attained and preserved. As he states: 'The subject of this Essay is not the so-called Liberty of the Will, but Civil, or Social Liberty: the nature and limits of the power which can be legitimately exercised by society over the individual.' He thus asserts a „very simple principle': 'that the sole end for which mankind are warranted, individually or collectively, in interfering with the liberty of action of any of their number, is self-protection. That the only purpose for which power can be rightfully exercised over any member of a civilized community, against his will, is to prevent harm to others. The only part of the conduct of any one, for which he is amenable to society, is that which concerns others. In the part which merely concerns himself, his independence is, of right, absolute. Over himself, over his own body and mind, the individual is sovereign.' One of his main trajectories is that in the course of democratization there is always the danger of tyranny of the majority. This is a danger one cannot be protected from, even though the majority might be wrong. In general, for Mill such liberty is only applicable and possible for people who are able to decide for themselves, therefore such notion excludes children and those he calls 'Barbarians.' John Stuart Mill develops a utilitarian ethical theory in Utilitarianism (1861/63). Once more he distinguishes between theories of intuition and those of experience, whereby it is the 'greatest happiness principle' that is the foundation of all utilitarian ethical principles. The ultimate goal of utilitarianism, Mill argues, is to achieve such happiness for the majority of people: 'To do as one would be done by, and to love one's neighbour as oneself, constitute the ideal perfection of utilitarian morality.'

John Stuart Mill's works have never shied away from controversy. Mill's Examination of Sir William Hamilton's Philosophy (1865) is said to contain the first secular appearance of Phenomenalism. It was only after this work that this doctrine became the standard among scientific philosophers during this time. Co-written with his wife, The Subjection of Women (1869), was quite radical for its time, and caused yet another controversy. The essay argued in favor of equality of the sexes employing utilitarian arguments. Mill was convinced that the inequality of women would impede human progress in general. For Mill the only way to find out whether there are actually differences between men and women, he argues, is by experiment. Mill chose to publish this essay relatively late in his life, possibly to avoid controversy. For similar reasons he did not want to have his Three Essays on Religion published during his lifetime (they only appeared in 1874), though these were not as radical as his followers had hoped. Convinced that they would impede progress, Mill here argues against an all-powerful god and traditional religious views in general. Instead, he proposes a 'religion of humanity,' wherein humanity itself is reverenced. During his lifetime, Mill was quite prolific and produced a great number of works in various fields such as economics, social philosophy, logic, but also religion and ethics. Among his most important ones are Essays on Some Unsettled Questions of Political Economy (1844), Thoughts on Parliamentary Reform (1859), Considerations on Representative Government (1861), Auguste Comte and Positivism (1865), England and Ireland (1869). His fragmented Autobiography was posthumously published in the same year of his death; he died on May 8, 1873 in Avignon, France.

4.4.1. Reciprocal Demand

Ricardo expounded the theory of comparative advantage without explaining the ratios at which commodities would exchange for one another. It was J. S. Mill who discussed the problem of ratios in detail in term of his theory of 'Reciprocal Demand'. The term 'reciprocal demand' was introduced by Mill to explain the determination of the equilibrium terms of trade. It is used to indicate a country's demand for one commodity in terms of the quantities of other commodities it is prepared to give up in exchange. It is reciprocal demand that determines the terms of trade which in turn determine the relative share of each country. Equilibrium would be established at that ratio of exchange between the two commodities at which quantities demanded by each country of the commodity which it imports from the other should be exactly sufficient to pay for another. J. S. Mill propounded the theory of reciprocal demand or the law of international values to explain the actual determination of equilibrium terms of trade. According to him, the equilibrium terms of trade are determined by the equation of reciprocal demand. Reciprocal demand means the relative strength and elasticity of demand of the two trading countries for each other's product in terms of their own product. A stable ratio of exchange will be determined at a level where the value of imports and exports of each country is in equilibrium. In Mill's own words, 'The actual ratio at which goods are traded will depend upon the strength and elasticity of each country's demand for the other country's product, or upon reciprocal demand. The ratio will be stable when the value of each country's exports is just enough to pay for its imports. To explain his theory of reciprocal demand, Mill first restated the Ricardian theory of comparative costs, 'Instead of taking as given the output of each commodity in two countries, with the labour costs different, he assumed a given amount of labour in each country but differing outputs. Thus his formation ran in terms of comparative advantage or comparative effectiveness of labour, as contrasted with Ricardo's comparative labour cost.

Assumptions

Mill's theory of reciprocal demand is based on the following assumptions.

1. There are two countries, say, England and Germany.
2. There are two commodities, say, linen and cloth.
3. Both the commodities are produced under the law of constant returns.
4. There are no transport costs.
5. The needs of the two countries are similar.
6. There is perfect competition.
7. There is full employment.
8. There is free trade between the two countries.
9. The principle of comparative costs is applicable in trade relations between the two countries. Miller's theory of reciprocal demand relates to the

possible terms of trade at which the two commodities will exchange for each other between the two countries. The terms of trade refer to the barter terms of trade between the two countries *i.e.* the ratio of the quantity of imports for given quantity of exports of a country. And 'the limits to the possible barter terms of trade (the international exchange ratio) are set by domestic exchange ratios established, by the relative efficiency of labour in each country.

4.4.2. Criticism of the Theory

Mill's theory of reciprocal demand is based on almost the same unrealistic assumptions that were adopted by Ricardo in his doctrine of comparative advantage. Thus the theory suffers from weaknesses. Besides, there are some additional criticism made by Viner, Graham and others.

- ☆ Mills theory of reciprocal demand does not take into account the domestic demand for the product, as pointed out by viner, each country would export its product only after satisfying its home demand. Thus the demand curve for Germany would not be below the line until the domestic demand was satisfied and same applies to England.
- ☆ According to Graham Mills analysis is valid only if the two countries are of equal size and the two commodities are of equal consumption value. In absence of these two assumptions if one country is small and the other large, the small country gains the most on both counts, first if it produced a high value commodity, it will adopt the cost ratios of its big partner and second the two trading countries being of unequal size the terms of trade will be fixed at or near the comparative costs of the large country. Graham further criticizes Mill for emphasizing demand and neglecting supply in determining international values. According to him the application of the reciprocal demand makes it appear that demand alone is of interest he maintains that production cost (supply) are also of paramount importance to international trade.
- ☆ Another weakness of Mill's reciprocal demand analysis is that it makes no allowance for fluctuations in incomes in the two trading countries which are bound to influence the terms of trade between them.
- ☆ Further the theory is based on barter terms of trade and relative price ratios. Thus it neglects all sickness of prices and wages, all transitional inflationary and over valuation gaps and all balance of payment problems.

Jeremy Bentham, the son of a lawyer, was born in London in 1748. A brilliant scholar, Bentham entered Queen's College, Oxford at twelve and was admitted to Lincoln's Inn at the age of fifteen. Bentham was a shy man who did not enjoy making public speeches. He therefore decided to leave Lincoln Inn and concentrate on writing. Provided with £90 a year by his father, Bentham produced a series of books on philosophy, economics and politics. Bentham's family had been Tories and for the first period of his life he shared their conservative political views. This changed after Bentham read the work of Joseph Priestley. One statement in particular from The First Principles of Government and the Nature of Political, Civil and Religious Liberty(1768) had a major impact on Bentham: 'The good and happiness of the members, that is the majority of the members of the state, is the great standard by which everything relating to that state must finally be determined.' Another important influence on Bentham was the philosopher David Hume. In books such as A Fragment on Government (1776) and Introduction to the Principles of Morals and Legislation (1789), Bentham argued that the proper objective of all conduct and legislation is 'the greatest happiness of the greatest number'. According to Bentham, 'pain and pleasure are the sovereign masters governing man's conduct'. As the motive of an act is always based on self-interest, it is the business of law and education to make the sanctions sufficiently painful in order to persuade the individual to subordinate his own happiness to that of the community.

In 1798 Bentham wrote Principles of International Law where he argued that universal peace could only be obtained by first achieving European Unity. He hoped that some form of European Parliament would be able to enforce the liberty of the press, free trade, the abandonment of all colonies and a reduction in the money being spent on armaments. In Catechism of Reformers (1809) Bentham criticized the law of libel as he believed it was so ambiguous that judges were able to use it in the interests of the government. Bentham pointed out that the authorities could use the law to punish any Radical for 'hurting the feelings' of the ruling class. Bentham also attacked other aspects of the legal system such as 'jury packing'. Radical reformers such as Sir Francis Burdett, Leigh Hunt, William Cobbett, and Henry Brougham praised Bentham's work. Although written in a complex style, radical publishers attempted to communicate his ideas to the working class. Jonathan Wooler published extracts in his journal Black Dwarf and eventually published a cheap edition of Catechism of Reformers. When Burdett introduced a series of resolutions in the House of Commons in July 1818, demanding universal suffrage, annual parliaments and vote by ballot, he quoted the writings of Jeremy Bentham to support his case.

In 1824 Bentham joined with James Mill (1773-1836) to found the Westminster Review, the journal of the philosophical radicals. Contributors to the journal included Lord Byron, Samuel Taylor Coleridge and Thomas Carlyle. Bentham's most detailed account of his ideas on political democracy appeared in his book Constitutional Code (1830). In the book Bentham argued that political reform should be dictated by the principal that the new system will promote the happiness of the majority of the people affected by it. Bentham argued in favour of universal suffrage, annual parliaments and vote by ballot. According to Bentham there should be no king, no House of Lords, no established church. The book also included Bentham's view that women, as well as men, should be given the vote. In Constitutional Code Bentham also addressed the problem of how government should be organized. For example, he suggested the introduction of rules that would ensure the regular attendance of members of the House of Commons. Government officials should be selected by competitive examination. The book also suggested the continual inspection of the work of politicians and government officials. Bentham pointed out they should be continually reminded that they are the 'servants, not the masters, of the public'. Jeremy Bentham died in 1832. Bentham's main contribution to the fields of philosophy and economics was that of Utilitarianism. Utilitarianism was the principle that the greatest happiness or utility was the most important, most valuable, etc., furthermore, such a belief led that people would strive to maximize their net pleasure (or, spoken alternatively: maximize their pleasure while minimizing their pain) - a strong origin for the theories of maximization. Utilitarianism is heavily studied in the field of ethical philosophy striving to answer the question somewhat addressed by Adam Smith: if everyone strives to maximize their personal utility, will the general benefit be maximized?

Bentham identified the fact that 'communities' are truly just the conglomeration of individuals, and thus, the interests of the 'community' are truly just the interest of the sum of the individuals. The ideas spoken with regard to hedonism and Utilitarianism often led Bentham to an important contribution to economics - that of diminishing utility. Adding an additional cart-and-buggy to an individual who already possesses dozens is going to provide very little utility, but an individual who doesn't have a cart-and-buggy may gain the benefit of being able to transport more, and otherwise increase their gain more. The effect of wealth in the production of happiness is diminishing, said Bentham. Diminishing marginal utility unquestionably supported a position of wealth redistribution when taken alone, Bentham supported this idea, saying that if you took an amount from someone with much money, they'd feel the loss less than someone who received it would feel the benefit. Bentham rejected modern ideas of GDP and other metrics representing good measures of wellbeing or happiness, saying that money was not a good metric of people's happiness. He called on politicians and moralists to discover a new metric of well-being, for the hopes of making morality an exact science were progressive. Bentham introduced the idea of compensation for labour ('wages') being simply payment for the fact labour is 'painful.'

Jeremy Bentham is most often associated with his theory of utilitarianism, the idea that all social actions should be evaluated by the axiom 'It is the greatest happiness of the greatest number that is the measure of right and wrong.' Counter to Adam Smith's vision of 'natural rights,' Bentham believed that there were no natural rights to be interfered with. Utilitarianism is mainly characterized by two elements: happiness and consequentialism. Utilitarian happiness is the biggest happiness which every human being looks for. In utilitarianism everything useful to happiness is good. Therefore, the name of the doctrine is utilitarianism, based on the principle of utility. Utility is found in everything which contributes to the happiness of every rational being. The criterion of good and evil is balanced between individual's happiness and the happiness of the community, 'each counting in an equal way' (Bentham, Introduction in the principles of morality and legislation). Consequentialism in utilitarianism is in the fact that an action must be judged for its consequences on the happiness of the largest number. That is: my search for happiness stops when it decreases the happiness of another individual or the happiness of the largest number, of the society or the community. As personal freedom is considered in respect of the freedom of other individuals and of the community, my freedom stops when it diminishes the freedom of another individual or the wellbeing of the society. We could say that utilitarianism is the continuation of Roman legislation, and its modern aspect is shown in the fact that utilitarianism adds an economical, legislative and political dimension to an ethical concept, that of happiness and well-being. The modern aspect of the doctrine will evolve throughout the 19th century, with Bentham, Mill and Sidgwick who succeeds in giving to this doctrine a practical and rational dimension which we can find in our modern society, in economics, politics and ethics.

'The continuing vitality of the greatest happiness system is not difficult to understand – it embodies a very natural and compelling model of rationality. This model, which dominates much of contemporary economics (as well as decision theory, 'cost-benefit analysis', and 'public choice theory')sees rational action as an attempt to maximise net utility (*i.e.* the result of summing the benefits and costs and subtracting the latter from the former). This view, which is frequently called 'means-end' rationality, goes back (at least) to Aristotle. In the Nicomachean Ethics Aristotle asserts that 'we cannot deliberate about ends but only about the means by which ends can be attained.' If we assume, with Aristotle, that happiness is the 'highest good attainable by action,' and hence the aim of politics, we get something very like Bentham's view. Indeed it is tempting, and not implausible, to interpret philosophers as different as Adam Smith and Chairman Mao as agreeing that the goal of social institutions is the maximization of realizing that end. Of course philosophers who share this vision of the proper function of social institutions like law and morality may differ on more than the best methods to attain it, as Aristotle noted, there is widespread agreement that happiness is the goal, but considerable disagreement as to what constitutes happiness. For Bentham the answer is simple: happiness is just pleasure and absence of pain. The value (or disvalue) of a pleasure (or pain) depends only on its intensity and duration, and can (at least in principle) be quantified precisely. Given this, we can reconstruct one line of Bentham's argument for the principle of Utility as something like the following:

- The good of a society is the sum of happiness of the individuals in that society.
- The purpose of morality is promotion of the good of society.
- A moral principle is ideal if and only if universal conformity to it would maximize the good of society.
- Universal conformity to the principle of Utility ('Act always so as to maximize total net balance of pleasures and pains') would maximize the good of society.

Utilitarianism gets its name from Bentham's test question, 'What is the use of it?' He conceived of the idea when he ran across the words 'the greatest happiness of the greatest number' in Joseph Priestly's Treatise of Government. Jeremy Bentham developed his ethical system around the idea of pleasure. He built it on ancient hedonism which pursued physical pleasure and avoided physical pain. According to Bentham, the most moral acts are those which maximize pleasure and minimize pain. This has sometimes been called the 'utilitarian calculus.' An act would be moral if it brings the greatest amount of pleasure and the least amount of pain. John Stuart Mill modified this philosophy and developed it apart from Bentham's hedonistic foundation. Mill used the same utilitarian calculus but instead focused on maximizing the general happiness by calculating the greatest good for the greatest number. While Bentham used the calculus in a quantitative sense, Mill used this calculus in a qualitative sense. He believed, for example, that some pleasures were of higher quality than others. Utilitarianism has been embraced by so many simply because it seems to make a good deal of sense and seems relatively simple to apply. However, when it was first proposed, utilitarianism was a radical philosophy. It attempted to set forth a moral system apart from divine revelation and biblical morality. Utilitarianism focused on results rather than rules. Ultimately the focus on the results demolished the rules. In other words, utilitarianism provided for a way for people to live moral lives apart from the Bible and its prescriptions. There was no need for an appeal to divine revelation. Reason rather than revelation was sufficient to determine morality.

There are a number of reasons for popularity of utilitarianism First, it is a relatively simple ethical system to apply. To determine whether an action is moral you merely have to calculate the good and bad consequences that will result from a particular action. If the good outweighs the bad, then the action is moral. Second, utilitarianism avoids the need to appeal to divine revelation. Many adherents to this ethical system are looking for a way to live a moral life apart from the Bible and a belief in God. The system replaces revelation with reason. Logic rather than an adherence to biblical principles guides the ethical decision-making of a utilitarian. Third, most people already use a form of utilitarianism in their daily decisions. We make lots of non-moral decisions every day based upon consequences. At the checkout line, we try to find the shortest line so we can get out the door more quickly. We make most of our financial decisions (writing checks, buying merchandise, etc.) on a utilitarian calculus of cost and benefits. So making moral decisions using utilitarianism seems like a natural extension of our daily decision-making procedures.

There are also a number of problems with utilitarianism. One problem with utilitarianism is that it leads to an 'end justifies the means' mentality. If any worthwhile end can justify the means to attain it, a true ethical foundation is lost. But we all know that the end does not justify the means. If that were so, then Hitler could justify the Holocaust because the end was to purify the human race. Stalin could justify his slaughter of millions because he was trying to achieve a communist utopia. The end never justifies the means. The means must justify themselves. A particular act cannot be judged as good simply because it may lead to a good consequence. The means must be judged by some objective and consistent standard of morality. Second, utilitarianism cannot protect the rights of minorities if the goal is the greatest good for the greatest number. Americans in the eighteenth century could justify slavery on the basis that it provided a good consequence for a majority of Americans. Certainly the majority benefited from cheap slave labour even though the lives of black slaves were much worse. A third problem with utilitarianism is predicting the consequences. If morality is based on results, then we would have to have omniscience in order to accurately predict the consequence of any action. But at best we can only guess at the future, and often these educated guesses are wrong. A fourth problem with utilitarianism is that consequences themselves must be judged. When results occur, we must still ask whether they are good or bad results. Utilitarianism provides no objective and consistent foundation to judge results because results are the mechanism used to judge the action itself.

4.5. Socialists

Socialism arose in the late 18th and early 19th century as a reaction to the economic and social changes associated with the Industrial Revolution. While rapid wealth came to the factory owners, the workers became increasingly impoverished. As this capitalist industrial system spread, reactions in the form of socialist thought increased proportionately. Although many thinkers in the past expressed ideas that were similar to later socialism, the first theorist who may properly be called socialist was François.

Noël Babeuf, who came to prominence during the French Revolution. Babeuf propounded the doctrine of class war between capital and labour later to be seen in Marxism. Socialist writers who followed Babeuf, however, were more moderate. Known as 'utopian socialists,' they included Saint-Simon, Charles Fourier, and Robert Owen. Saint-Simon proposed that production and distribution be carried out by the state. The leaders of society would be industrialists who would found a national community based upon cooperation and who would eliminate the poverty of the lowest classes. Fourier and Owen, though differing in many respects, both believed that social organization should be based on small local collective communities rather than the large centralist state of Saint-Simon. All these men agreed, however, that there should be cooperation rather than competition and they implicitly rejected class struggle. In the early 19th cent. Numerous utopian communistic settlements founded on the principles of Fourier and Owen sprang up in Europe and the United States; New Harmony and Brook Farm were notable examples. Following the utopians came thinkers such as Louis Blanc who were more political in their socialist formulations. Blanc put forward a system of social workshops (1840) that would be controlled by the workers themselves with the support of the state. Capitalists would be welcome in this venture, and each person would receive goods in proportion to his or her needs. Blanc became a member of the French provisional government of 1848 and attempted to put some of his proposals into effect, but his efforts were sabotaged by his opponents. The anarchist Pierre Joseph Proudhon and the insurrectionist AugusteBlanqui were also influential socialist leaders of the early and mid-19th cent. The term 'Socialism' made its appearance in print in England in 1827. Five years later, the term was used for the first time in a French publication. It is no accident that the socialist idea –and the socialist movement–first appeared in England and France. For socialism was a product of two revolutions in human affairs, each with their respective roots in those two countries: the industrial revolution in England and the popular-democratic revolution in France.

The French revolution of 1789-1799 involved the most massive popular struggles that had yet been seen in history. Rooted in popular hatred of an oppressive monarchy, the revolution rose on the backs of the masses of poor people in Paris who united under the banner of 'liberty, equality and brotherhood'. Beginning as a rebellion against the abuses of the monarchy, the revolution grew into a massive challenge to all forms of oppressive authority– whether it was that of lords, priests or factory owners. Initially, the battle against the monarchy unified large sections of society. As the revolution advanced, however, a new ruling group tried to halt the process in order to maintain their grossly unequal system of property and power. As a result, the popular movement divided into conservative and revolutionary camps. In the conservative camp were those who saw freedom simply in terms of the freedom to own property. In the revolutionary camp were those who represented the Paris poor and who recognised that freedom was impossible without equality; that it was meaningless to talk of liberty if this was confined to the right of some men and women to starve to death while others grew rich off the labour of others. Out of the French Revolution, then, emerged the essential socialist idea that democracy and freedom require a society of equality. The French radicals recognised that genuine freedom presupposed the liberty of all to participate equally in producing and sharing the wealth of society. They understood that if some had the unequal right to own and monopolise land, wealth or factories, then others might just as unequally be condemned to a life of drudgery, misery and poverty. But a society of equality requires a state of abundance. So long as economic life remains relatively backward, equality can only mean the common hardship of shared poverty. A healthy and thriving popular democracy requires a state of prosperity in which all the basic needs of people can be satisfied. Without a certain level of economic development, therefore, the French revolutionaries' demand for liberty and equality remained utopian. It was only with the enormous economic development unleashed by the industrial revolution in England that a society based upon equality and abundance became a realistic possibility. The English industrial revolution conjures up images of dark and dirty textile mills, of ten-yearold children labouring in coal mines, of women and men working 12 and 14-hour days–in short, of suffering and misery. Such an image is largely correct. The industrial revolution that swept Britain, beginning in the last quarter of the eighteenth century, was a massive dislocation in social life: old communities were destroyed; people were forced off the land and into the tyranny of the factory; industrial diseases multiplied; hunger, poverty and illness spread; life expectancy fell. At the same time, however, several ingredients of the industrial revolution held out the prospect of an end to these ills. The new machinery of production that developed, especially during the early 1800s, offered the possibility of sharply reducing drudgery and toil and of massively increasing the production of wealth so as to eliminate poverty forever. In reality, the industrial revolution did no such thing. Rather than leading to an improvement in the conditions of labour, the new industry was used to increase the fortunes of a few–the new industrial capitalists. Nonetheless, some writers saw in the industrial revolution an enormous potential for improving the human condition. Even some well- intentioned bankers and factory owners came to believe that the forces of the industrial revolution should be harnessed to serve human ends. Many of these become early advocates of what has become known as 'utopian socialism'.

Britain's best known utopian socialist was the cotton manufacturer Robert Owen. Like most of the early socialists drawn from the capitalist class, Owen did not call for a mass, democratic restructuring of society. For Owen, the working class was a pathetic and pitiful group. Owen's socialism was based on appealing to wealthy leaders of business and government in order to persuade them to improve the wretched conditions of the labouring masses. In this respect, Owen was similar to the two earliest French utopian socialists, Henri Saint-Simon and Charles Fourier. Saint-Simon was a real estate speculator turned banker who rose to great wealth in the decades after the French Revolution. Fascinated by the enormous potential of science and technology, Saint-Simon began to argue the case for a 'socialist' society that would eliminate the disorderly aspects of capitalism. Saint-Simon's 'socialism' was decidedly anti-democratic. He did not envisage an expansion of human rights and freedoms. Instead, he hoped for a planned and modernised industrial society ruled over by an international committee of bankers. In many respects, Saint-Simon anticipated the development of state capitalism; he looked forward to a capitalist system in which industry would be owned and directed by a government made up of a scientists, managers and financiers. The socialism of Charles Fourier had more to commend it. A self-taught eccentric, Fourier developed some highly original ideas. But Fourier's outlook suffered from two main defects. First, he dismissed the potential of modern industry for bringing into being a society of abundance and looked nostalgically for a return to preindustrial conditions of life. Secondly, Fourier looked not to the masses of working people but to enlightened rulers to usher in the socialist utopia. He spent his time drawing up rigid blueprints for the new society and sent copies to rulers like the Czar of Russia and the President of the United States.

Indeed, this is the common thread that runs through the outlook of all the early utopian socialists. Each of them looked to some well- intentioned members of the ruling class to bring about a socialist transformation of society. Each rejected the notion that socialism could only be achieved democratically–through the mass action of working people. For this reason, all their views car be described as variants of socialism from above–a view in which the masses of people are mere playthings of an enlightened elite who will change society in the interests of the masses of people. There was, however, one revolutionary doctrine of socialism during this period. This consisted of what can best be called conspiratorial communism. Out of the defeat of the popular struggles of the French Revolution, one far-sighted group of rebel centred around a man named Gracchus Babeuf, developed a communist perspective. Babeuf and his followers believed that true democracy could only be constructed on the basis of common ownership of wealth. But they could see no way of winning a majority of society to support their communist programme. The masses of French people sought little else than protection of their own private property–their plot of land or their workshop. They showed little interest in a socialist transformation of society. For this reason, Babeuf–and his later follower, AdolpheBlanqui- could only conceive of a revolution made by a minority, the communist elite. As a result, democracy remained foreign to their socialist programme as well.

The origins of socialism as a political movement lie in the Industrial Revolution. Its intellectual roots, however, reach back almost as far as recorded thought–even as far as Moses, according to one history of the subject. Socialist or communist ideas certainly play an important part in the ideas of the ancient Greek philosopher Plato, whose'Republic' depicts an austere society in which men and women of the 'guardian' class share with each other not only their few material goods but also their spouses and children. Early Christian communities also practiced the sharing of goods and labour, a simple form of socialism subsequently followed in certain forms of monasticism. Several monastic orders continue these practices today. Christianity and Platonism were combined in More's Utopia, which apparently recommends communal ownership as a way of controlling the sins of pride, envy, and greed. Land and houses are common property on More's imaginary island of Utopia, where everyone works for at least two years on the communal farms and people change houses every 10 years so that no one develops pride of possession. Money has been abolished, and people are free to take what they need from common storehouses. All the Utopians live simply, moreover, so that they are able to meet their needs with only a few hours of work a day, leaving the rest for leisure. More's Utopia is not so much a blueprint for a socialist society as it is a commentary on the failings he perceived in the supposedly Christian societies of his day. Religious and political turmoil, however, soon inspired others to try to put utopian ideas into practice. Common ownership was one of the aims of the brief Anabaptist regime in the Westphalia city of Monster during the Protestant Reformation, and several communist or socialist sects sprang up in England in the wake of the Civil Wars (1642–51). Chief among them was the Diggers, whose members claimed that God had created the world for people to share, not to divide and exploit for private profit. When they acted on this belief by digging and planting on land that was not legally theirs, they ran afoul of Oliver Cromwell's Protectorate, which forcibly disbanded them.Whether utopian or practical, these early visions of socialism were largely agrarian. This remained true as late as the French Revolution, when the journalist François-Noël Babeuf and other radicals complained that the Revolution had failed to fulfill the ideals of liberty, equality, and fraternity. Adherence to 'the precious principle of equality,' Babeuf argued, requires the abolition of private property and common enjoyment of the land and its fruits. Such beliefs led to his execution for conspiring to overthrow the government. The publicity that followed his trial and death, however, made him a hero to many in the 19th century who reacted against the emergence of industrial capitalism.

Robert Owen was born in Newtown, Montgomeryshire (Wales) on May 14, 1771, the sixth of seven children. His father was a saddler and ironmonger who also served as local postmaster; his mother came from one of the prosperous farming families of Newtown.Owen attended the local school where he developed a strong passion for reading. At the age of ten hewas sent to seek his fortune in London with his eldest brother, William. After a few weeks, Owen found aposition in a large drapery business in Stamford (Lincolnshire) where he served as an apprentice. Afterthree years he returned to London where he served under another draper. Then, in 1787 or 1788, hemoved to Manchester in the employ of Mr. Satterfield, a wholesale and retail drapery merchant. Owen now found himself in what would soon become the capital city of the English Industrial Revolution on the eve of that event as factories were built and textile manufacture expanded. He was a serious, methodical young man who already possessed an extensive knowledge of the retail aspect of his chosen trade. In late 1790 he borrowed £100 from his brother William and set up independently with a mechanic named Jones as a manufacturer of the new spinning mules. After a few months he parted with Jones and started business on his own with three mules as a cotton spinner. During 1792, Owen applied for and was appointed manager of Peter Drinkwater's new spinning factory, the Piccadilly Mill, where he quickly achieved the reputation as a spinner of fine yarns, thanks to the application of steam power to the mule. One of Drinkwater's most important a client was Samuel Old know, maker of fine muslins. Drinkwater had intended Owen to become a partner in his new business by 1795, but a projected marriage alliance between Drinkwater's daughter and Old know caused the cancellation of the agreement with Owen. Hurt and unwilling to remain a mere manager, Owen left Piccadilly Mill in 1795.

Owen was approached by Samuel Marsland who intended to develop the Charlton estate in Manchester, but instead he found partners in two young and inexperienced businessmen, Jonathan Scarth and Richard Moulson, who undertook to erect cotton mills on land bought from Marsland, and the three partners were assisted by Marsland. In 1796, the financial basis of the company was broadened with the inclusion of Thomas Atkinson, thus constituting the Charlton Twist Company, which in 1799 negotiated the purchase of David Dale's New Lanark mills. The industrial community at New Lanark had been planned by Richard Arkwright and David Dale in 1783, to take advantage of the water power of the Falls of Clyde deep in the river valley below the burgh of Lanark, twenty-four miles upstream from of Glasgow. In 1800, there were four mills making New Lanark the largest cotton-spinning complex in Britain, and the population of the village (over 2000) was greater than that of Lanark itself. Dale was progressive both as a manufacturer and as an employer, being especially careful to safeguard the welfare of the children.

Owen first met Dale by chance through an introduction by the daughter of his friend, Robert Spear, to Dale's eldest daughter, Caroline. Owen was interested to learn that Dale wanted to sell New Lanark to someone who would continue his humane policy toward the children. Owen's willingness to do so was probably responsible for both Dale's agreeing to sell to the Chorlton Twist Company and also his consent to the marriage of Owen and Caroline in the fall of 1799. New Lanark made Owen's reputation as a philanthropist. The village remained much as Dale had made it although more living space was created and higher standards of hygiene were enforced. The primary contribution of Owen at new Lanark was in public buildings which emphasized his concern for the welfare of his workers, specifically, the New Institution for the Formation of Character (1816), the Infant School (1817) and the Store. Owen's partners did not share his enthusiasm for education and welfare: the major expenditure on social buildings came only after the formation of his third partnership in 1813. His ideas were shaped by the Enlightenment, his contact with progressive ideas in Manchester as a member of the Literary and Philosophical Society, and his acquaintance with the ideas of the Scottish Enlightenment. Owen's general theory was that character is formed by the effects of the environment upon the individual. Hence, education was of central importance to the creation of rational and humane character, and the duty of the educator was to provide the wholesome environment, both mental and physical, in which the child could develop. Physical punishment was prohibited and child labour was restricted. Man, being naturally good, could grow and flourish when evil was removed. Education, as one historian has put it, was to the 'the steam engine of his new moral world.' At New Lanark, Owen was responsible for overall management and general policy. His employees did not at first enjoy his attempts to regulate and improve their lives and his paternalism was more rigorous than his frequently absent partner, Dale. As a successful manager of people and business, Owen displayed a skill well in advance of his day but his welfarism, which was not really that unique, had a practical side. His store helped to raise real wages and the infant school enabled mothers to return to work when their children reached the age of one year. As a financier, Owen's methods were sophisticated. He had relatively little capital of his own, yet his skilful management of partnerships enabled him to survive and become a wealthy man before leaving New Lanark in 1828. At New Lanark, Owen involved himself in the public affairs of the day, the most important being education, factory reform, and the improvement of the Poor Laws. His first public speech was on education (1812) and was elaborated upon in his first published work, The First Essay on the Principle of the Formation of Character (1813). Together with three further essays (1813-1814), this comprised A New View of Society, which remains Owen's clearest declaration of principles. Owen's explicit denunciation of religion evoked a mounting campaign against him which in later years damaged his public reputation and the work associated with his name. His last real opportunity to secure official approval for his scheme came in 1820 when he produced his Report to the County of Lanark in which his communitarian and educational theories were blended with David Ricardo's labour theory of value.

By the late 1820s, Owen's roots in New Lanark were loosening. Owen now set about his mission to bring about the new moral world through his plan for well-regulated communities. England, Scotland and Ireland seemed indifferent, but the United States opened up new prospects and in 1824 Owen crossed the Atlantic and viewed the Rappite community at Harmony (Indiana), which was for sale. He bought it for£30,000 and in April 1825, initiated New Harmony. With Robert Dale left in charge of New Lanark and William Owen at New Harmony, their father travelled between the two, collecting his 'boatload of knowledge,' which reached the community in January 1826. The New Harmony community was not a success. By May 1827, there were ten different sub-communities on the estate, and a year later failure was apparent. During his absence at New Harmony, the nature of Owen's support in England had begun to change. Working men were now listening to his message, democratic socialist ideas were being developed by men like William Thompson of Cork, and cooperative, labour exchange and trades union movements were becoming more popular. Owen became convinced that the world of competitive industrial capitalism had reached a stage of crisis and that the leaders of society would now turn to him in their hour of need. What Owen offered the working class Owenites was social salvation – his creed was that of the secular millennium. These views were expressed in his weekly periodical, The Crisis (1832-1834), and had a following particularly among the labour aristocrats of London who sought to exchange their products according to the labour theory of value at the Gray's Inn Road Labour Exchange, which Owen opened in 1832. Breaking with these labour movements in 1834, Owen turned back to his plan for a community and founded a journal, The New Moral World (November, 1834) and an organization, the Association of All Classes of All Nations (May, 1835) to prepare public opinion for the millennium.

In the 1840s, Owen embarked on a new settlement at Queenwood Farm in Hampshire. There was insufficient capital and the community, projected to support 500 members, never attracted more than ninety communitarians. In 1841, Owen secured capital from a consortium of capitalist friends and built a luxurious mansion, Harmony Hall, to house a community 'normal school' which would train Owenites in a correct communitarian environment. Owen quickly spent his funds and in July 1842 was removed from control. He resumed control in May 1843, but his concept of a 'normal school' was not what many Owenites had hoped for, and in 1844 the annual Owenite Congress rebelled against his despotic control of community policy. Owen moved on. His missions to Europe and North America never ceased and he retained a lively interest in current affairs, confidently expecting that governments would secure his services. In 1855 he called a series of public meetings to proclaim the millennium. A loyal nucleus of Owenites stood by his side, devoted to the man who, whatever else he had done, had given them a vision of a new moral world. In 1853 he became a spiritualist. On November 17, 1858 Owen died in the Bear Hotel, next door to the house in which he was born. Owen's character was a paradox to his contemporaries. By temperament, he was conservative and authoritarian; by nature he was naive. He was convinced that man's character was made for him, rather than by him and that social change would only come from calm reasoning with the leaders of society. He never believed in the independent power of the working classes and he could never conceive that within capitalist society there might be more than one rationally agreed interest.

Robert Owen is considered the father of the cooperative movement. Robert Owen believed in putting his workers in a good environment with access to education for themselves and their children. These ideas were put into effect successfully in the cotton mills of New Lanark, Scotland. It was here that the first co-operative store was opened. Spurred on by the success of this, he had the idea of forming 'villages of co-operation' where workers would drag themselves out of poverty by growing their own food, making their own clothes and ultimately becoming self-governing. He tried to form such communities in Orbiston in Scotland and in New Harmony, Indiana in the United States of America, but both communities failed.

Charles Fourier was born in Besancon, France, on 7th April, 1773. The son of a cloth merchant, he was educated at the local Jesuit College. After serving the French Army he worked as a clerk in Lyon. In 1808 he published his first book, The Social Destiny of Man. In the book Fourier criticized the immorality of the business world, arguing that 'truth and commerce are as incompatible as Jesus and Satan.' In the book Fourier advocated a new socialist system of cooperation. He suggested that 'phalanxes' should be established. These would be scientifically planned to offer a maximum of both cooperation and self-fulfilment to their members. Fourier suggested that these communes should contain about 1,600 people and should attempt to be compatible with each member's 'natural talents, passions, and inclinations'.The ideas in the book influenced writers such as Alexander Hearken, Peter Larva, Pierre-JosephProudhon, Prince Kropotkin, Ralph Waldo Emerson and Henry David Thoreau. Others like Karl Marx andFrederick Engels accused Fourier of being 'utopian' and attempted to develop a more scientific theory of socialism. One of Fourier's supporters, Victor Considerant, established a newspaper in order to promote the cause. Others attempted to establish their own *phalanstery*. This included one at Rambouillet in France that was under-capitalized and eventually went bankrupt (1834-36). Another, more successful attempt, was made by George Ripley at Brook Farm in Massachusetts (1841-46).Although no long-term phalanxes were established, Fourier's ideas influenced a generation of socialists, anarchists, feminists, pacifists, internationalists and others questioning the morality of the capitalist system. Even Karl Marx and Frederick Engels used Fourier's ideas to develop their theory of alienation. Fourier also published The New World of Communal Activity (1829) and The False Division of Labour (1835). However, his attempts to find a rich benefactor to fund a *phalanstery* ended in failure.Charles Fourier died in Paris on 10th October, 1837. The phalange, in Fourier's conception, was to be a cooperative agricultural community bearing responsibility for the social welfare of the individual, characterized by continual shifting of roles among its members. He felt that phalanges would distribute wealth more equitably than under capitalism and that they could be introduced into any political system, including a monarchy. The individual member of a phalange was to be rewarded on the basis of the total productivity of the phalange.

Saint-Simon, was born in Paris, France, in 1760 as the son of a minor noble. Privately educated he served in the French Army during the American War of Independence. Afterwards he travelled to Mexico and Spain where he became involved in several canal projects. A supporter of the French Revolution, he immediately renounced his title. He was imprisoned during the Terror but was released after spending nine-months in captivity. His first book on political theory, Letters of a Geneva to His Contemporaries, was published in 1802. This was followed by Introduction to the Work of Science in the 19th Century (1807), Memoir on the Science of Man (1813), On the Reorganization of European Society (1814), and The New Christianity (1825). In his books Saint-Simon argued that Europe was in 'critical disequilibrium' and would soon undergo reconstruction. He argued strongly for a planned economy. He suggested a framework of three chambers: one body made up of engineers and artists to propose plans, a second of scientists responsible for assessing the plans, and a third group of industrialists whose task would be that of implementing the schemes according to the interests of the whole community. After his death in 1825, Saint-Simon's ideas were developed by a group of loyal followers such as Olindes Rodriguez, Armand Bazard and Barthelemy-Prosper Enfantin. In 1830 the group published An Explanation of the Doctrine of Saint-Simon. The interpreted Saint-Simon as being a socialist and argued for the redistribution of wealth for the benefit of society. Saint-Simon's theories also influenced figures such as Alexander Herzen, Thomas Carlyle and J. S. Mill.

Saint-Simon was a nobleman who led a somewhat dissolute, adventurouslife. At the early age of sixteen he took part in the American War of Independence. The Revolution witnessed the abandonment of his claim to nobility, but by successful speculation in national property he was enabled to retrieve his fortune to some extent. Imprisoned as a suspect at Sainte-Pelagie, set free on the 9th Thermidor, he attained a certain notoriety as a man of affairs interested chiefly in travels and amusements and as a dilettante student of the sciences. From the moment of his release he began to regard himself as a kind of Messiah. He was profoundly impressed by what seemed to him to be the birth of a new society at which he had himself assisted, in which the moral and political and even physical conditions of life were suddenly torn up by the roots, when ancient beliefs disappeared and nothing seemed ready to take their place. He himself was to be the evangelist of the new gospel, he called together the capitalists who were already associated with him and, pointing out the great necessity for restoring public confidence, proposed the establishment of a gigantic bank whose funds might be employed in setting up works of public utility-proof of the curious way in which economic and philosophic considerations were already linked together in his thoughts. An ill-considered marriage which was hastily broken off, however, was followed by a period of much extravagance and great misery. By the year 1805 so reduced were his circumstances that he was glad to avail himself of the generosity of one of his old servants. After her death he lived partly upon the modest pension provided him by his family and partly upon the contributions of a few tradesmen, but he was again so miserable that in 1823 he attempted suicide. A banker of the name of Olinde Rodrigues came to the rescue this time and supplied him with the necessary means of support. He died in 1825, surrounded by a number of his disciples who had watched over the last moments of his earthly life. During all these years, haunted as he was by the need for giving to the new century the doctrine it so much required, he was constantly engaged in publishing brochures, new works, or selections from his earlier publications, sometimes alone and sometimes in collaboration with others, in which the same suggestions are always revived and the same ideas keep recurring, but in slightly different forms. Saint-Simon's earlier work was an attempt to establish a scientific synthesis which might furnish mankind with a system of positive morality to take the place of religious dogmas. It was to be a kind of 'scientific breviary' where all phenomena could be deduced from one single idea, that of 'universal gravitation.' He himself has treated us to a full account of this system, which is as deceptive as it is simple, and which shows us his serious limitations as a philosopher whose ambition far outran his knowledge. AugusteGomte, one of his disciples, attempted a similar task in his *Cours de Philosophic* positive and in the *Politique* positive, so that Saint-Simon, who is usually considered the father of socialism, finds himself also the father of positivism. From 1814 up to his death in 1825 he partly relinquished his interest in philosophy and devoted himself almost exclusively to the exposition of his social and political ideas, which are the only ones that interest us here.

His economics might be summed up as an apotheosis of industry, using the latter word in the widest sense, much as Smith had employed the term as synonymous with labour of all kinds.

His leading ideas, contained within the compass of a few striking pages, have since become known as 'Saint-Simon's Parable.' 'Let us suppose,' says he, that France suddenly loses fifty of her first-class doctors, fifty first-class chemists, fifty first-class physiologists, fifty first-class bankers two hundred of her best merchants, six hundreds of her foremost agriculturists, five hundred of her most capable ironmasters, etc. Seeing that these men are its most indispensable producers, makers of its most important products, the minute that it loses these the nation will degenerate into a mere soulless body and fall into a state of despicable weakness in the eyes of rival nations, and will remain in this subordinate position so long as the loss remains and their places are vacant. Let us take another supposition. Imagine that France retains all her men of genius, whether in the arts and sciences or in the crafts and industries, but has the misfortune to lose on the same day the king's brother, the Duke of Angouleme, and all the other members of the royal family; all the great officers of the Grown; all ministers of State, whether at the head of a department or not; all the Privy Councillors; all the masters of requests; all the marshals, cardinals, archbishops, bishops, grand vicars and canons; all prefects and sub-prefects; all Government employees; all the judges; and on top of that a hundred thousand proprietors- the cream of her nobility. Such an overwhelming catastrophe would certainly aggrieve the French, for they are a kindly-disposed nation. But the loss of a hundred and thirty thousand of the best reputed individuals in the State would give rise to sorrow of a purely sentimental kind. It would not cause the community the least inconvenience.

In other words, the official Government is a mere facade. Its action is wholly superficial. Society might exist without it and life would be none the less happy. But the disappearance of the savants, industrial leaders, bankers, and merchants would leave the community crippled. The very sources of wealth would dry up, for their activities are really fruitful and necessary. They are the true governors who wield real power. Such was the parable. According to Saint-Simon, little observation is needed to realize that the world we live in is based upon industry, and that anything besides industry is scarcely worth the attention of thinking people. A long process of historical evolution, which according to Saint- Simon commenced in the twelfth century with the enfranchisement of the communes and culminated in the French Revolution, had prepared the way for it. At least industry is the one cardinal feature of the present day.

The political concerns of his contemporaries were regarded with some measure of despair. The majority of them were engaged either in defending or attacking the Charter of 1814. The Liberals were simply deceiving themselves, examining old and meaningless formulae such as 'the sovereignty of the people,' liberty,' and 'equality'- conceptions that never had any meaning, but were simply metaphysical creations of the jurists, and they ought to have realized that this kind of work was perfectly useless now that the feudal regime was overthrown. Men in future will have something better to do than to defend the Charter against the 'ultras.' The parliamentary regime may be very necessary, but it is just a passing phase between the feudalism of yesterday and the new order of to-morrow. That future order is Industrialism- a social organization having only one end in view, the further development of industry, the source of all wealth and prosperity.

The new regime implies first of all the abolition of all class distinction. There will be no need for nobles, bourgeois, or clergy. There will be only two categories, workers and idlers–or the bees and the drones, as Saint-Simon puts it. Sometimes he refers to them as the national and anti-national party. In the new society the second class is bound to disappear, for there is only room for the first. This class includes, besides manual workers, agriculturists, artisans, manufacturers, bankers, savants, and artists. Between these persons there ought to be no difference except that which results from their different capacities, or what Saint-Simon calls their varying stakes in the national interest. 'Industrial equality,' he writes, 'consists in each drawing from society benefits exactly proportionate to his share in the State -that is, in proportion to his potential capacity and the use which he makes of the means at his disposal- including, of course, his capital.' Saint-Simon evidently has no desire to rob the capitalists of their revenues; his hostility is reserved for the landed proprietors. Not only must every social distinction other than that founded upon labour and ability disappear, but government in the ordinary sense of the term will largely become unnecessary. 'National association' for Saint-Simon merely meant 'industrial enterprise.' France was to be turned into a factory and the nation organized on the model of a vast workshop'; but 'the task of preventing thefts and of checking other disorders in a factory is a matter of quite secondary importance and can be discharged by subordinates.' In a similar fashion, the function of government in industrial society must be limited to 'defending workers from the unproductive sluggard and maintaining security and freedom for the producer.'

So far Saint-Simon's 'industrialism' is scarcely distinguishable from the 'Liberalism' of Smith and his followers, especially J. B. Say's. Charles Comte and Dunoyer, writing in their review, *Le Censeur,* were advancing exactly similar doctrines, sometimes even using identical terms. 'Plenty of scope for talent' and laissez-faire were some of the favourite maxims of the Liberal bourgeois. Such also were the aspirations of Saint-Simon. But it is just here that the tone changes. Assuming that France has become a huge factory, the most important task that awaits the nation is to inaugurate the new manufacturing regime and to seek to combine the interests of the entrepreneurs with those of the workers on the one hand and of the consumers on the other. There is thus just enough room for government -of a kind.

What is required is the organizing of forces rather than the governing of men. Politics need not disappear altogether, but 'must be transformed into a positive science of productive organization.' Under the old system the tendency was to increase the power of government by establishing the ascendancy of the higher classes over the lower. Under the new system the aim must be to combine all the forces of society in such a fashion as to secure the successful execution of all those works which tend to improve the lot of its members either morally or physically. Such will be the task of the new government, where capacity will replace power and direction will take the place of command. Applying itself to the execution of those tasks upon which there is complete unanimity, most of them requiring some degree of deliberation and yet promptness of action, it will gradually transform the character of politics by concentrating attention upon matters affecting life or wellbeing -the only things it need ever concern itself with.

Utopian socialism is a term used to define the first currents of modern socialist thought as exemplified by the work of Henri de Saint-Simon, Charles Fourier, and Robert Owen. The utopian socialist thinkers did not use the term utopian to refer to their ideas. Karl Marx and Friedrich Engels referred to all socialist ideas that were simply a vision and distant goal for society as utopian. Utopian socialists were likened to scientists who drew up elaborate designs and concepts for creating what socialists considered a more equal society. They were contrasted by scientific socialists, likened to engineers, who were defined as an integrated conception of the goal, the means to producing it, and the way that those means will inevitably be produced through examining social and economic phenomena. Utopian Socialism started in the early 1800's when men such as Robert Owen of Great Britain and Charles Fourier of France made proposals for setting up communities with ideal economic and social conditions. These socialists desired for the people to work together under one government to obtain perfect living conditions for everyone. The gained the name 'Utopians' from the book Utopia, written by the English statesman Saint Thomas Moore in 1516. The book spoke of an ideal society where everyone possessed equality. Utopians desired a classless society where everyone possessed the same amount of power under the government. They believed everyone deserved a decent house to live in and food on the table. No one owned private property because it all belonged to the public. While they wanted material equality for all men, the amount of work required from each man differed. While some performed much manual labour, others performed small tasks. Because Utopians required each man to give up his profits as public property, they destroyed the work ethic. Man made no personal gain, so he felt no need to work. This also stunted the invention of new items. It pulled man's desire to better himself to a halt because, ideally, he received all he needed from the government. Utopians seemed to live in a fantasy world with equality, rather than a realistic world where men desire individual power.

4.6. Important Utopian Socialists

Perhaps the first utopian socialist was Thomas More (1478-1535), who wrote about an imaginary socialist society in his satire Utopia, which was published in 1516. The contemporary definition of the English word 'utopia' derives from this work.Utopia – literally 'nowhere' – was the name of an imaginary republic described by Thomas More in which all social conflict and distress has been overcome. There have been many versions of Utopia over the years, many of them visions of socialist society. Although Marx and Engels defined their own socialism in opposition to Utopian Socialism (which had many advocates in the early nineteenth century), they had immense respect for the great Utopian socialists like Charles Fourier and Robert Owen.By describing how people would live if everyone adhered to the socialist ethic, utopian socialism does three things: it inspires the oppressed to struggle and sacrifice for a better life, it gives a clear meaning to the aim of socialism, and it demonstrates how socialism is ethical, that is, that the precepts of socialism can be applied without excluding or exploiting anyone. Saint-Simonwas a French political and social movement of the first half of the 19th century, inspired by the ideas of Claude Henri de Rouvroy, comte de Saint-Simon (1760–1825). His ideas influenced Auguste Comte (who was, for a time, Saint-Simon's secretary), Karl Marx, John Stuart Mill, and many other thinkers and social theorists. Robert Owen (1771–1858) was a successful Welsh businessman who devoted much of his profits to improving the lives of his employees. His reputation grew when he set up a textile factory in New Lanark, Scotland, co-funded by his teacher, the utilitarian Jeremy Bentham, and introduced shorter working hours, schools for children and renovated housing. He wrote about his ideas in his book A New View of Society, which was published in 1813, and An Explanation of the Cause of Distress which pervades the civilized parts of the world in 1823. He also set up an Owenite commune called New Harmony in Indiana, USA. This collapsed when one of his business partners ran off with all the profits. Owen's main contribution to socialist thought was the view that human social behaviour is not fixed or absolute, and that human beings have the free will to organize themselves into any kind of society they wished.

Charles Fourier (1772–1837) was by far the most utopian of the socialists. Rejecting the industrial revolution altogether and thus the problems that arose with it, he made various fanciful claims about the ideal world he envisioned. Despite some clearly non-socialist inclinations, he contributed significantly - if indirectly - to the socialist movement. His writings about turning work into play influenced the young Karl Marx and helped him devise his theory of alienation. Also a contributor to feminism, Fourier invented the concept of *phalanstère*, units of people based on a theory of passions and of their combination. Several colonies based on Fourier's ideas were founded in the United States by Albert Brisbane and Horace Greeley. Étienne Cabet (1788–1856) who was influenced by Robert Owen, published a book in 1840 entitled Travel and adventures of Lord William Carisdall in Icaria in which he described an ideal communalist society. His attempts to form real socialist communities based on his ideas, through the Icarian movement however, did not survive, but one such community was the precursor of Corning, Iowa. Possibly inspired by Christianity, he coined the word 'communism' and influenced other thinkers, including Karl Marx and Friedrich Engels. Edward Bellamy (1850–1898), published Looking Backward in 1888, a utopian romance novel about a future socialist society. In Bellamy's utopia, property was held in common and money replaced with a system of equal credit for all. Valid for a year and non-transferable between individual persons, expenditure of this credit was to be tracked via 'credit-cards'. Labour between the ages of 21-40 was to be compulsory, and organized via various departments of an 'Industrial Army' to which most citizens belonged. However working hours were to be cut drastically due to technological advances. People were expected to be motivated by a Religion of Solidarity, and *e.g.* criminal behaviour was treated as a form of mental illness or 'atavism'. It was the second or third ranking best seller of its time.

Bellamy published a sequel Equality in 1897 as a reply to his critics, and from which the Industrial Army and other authoritarian aspects were absent. William Morris (1834–1896) published News from Nowhere in 1890, partly as a response to Bellamy's Looking Backwards which he equated with the socialism of Fabians such as Sydney Webb. Morris' vision of the future socialist society was centered on his concept of useful work as opposed to useless toil, and the redemption of human labour. Morris believed that all work should be artistic, in the sense that the worker should find it both pleasurable and an outlet for creativity. Morris' conception of labour thus bears strong resemblance to Fourier's, whilst Bellamy's (*i.e.* the reduction of labour to a minimum) is more akin to that of Saint-Simon or indeed Marx.

5

Marxian Political Economy

5.1. Marxian Political Economy

The political and economic philosophy of Karl Marx and Friedrich Engels in which the concept of class struggle plays a central role in understanding society's allegedly inevitable development from bourgeois oppression under capitalism to a socialist and ultimately classless society is generally referred to as Marxian Political Economy model. Political economy was the original term used for studying production, buying, and selling, and their relations with law, custom, and government, as well as withthe distribution of national income and wealth. Political economy originated in moral philosophy. It was developed in the 18th century as the study of the economies of states, or polities, hence the term political economy. In the late 19th century, the term economics came to replace political economy, coinciding with the publication of an influential textbook by Alfred Marshall in 1890. Earlier, William Stanley Jevons, a proponent of mathematical methods applied to the subject, advocated economics for brevity and with the hope of the term becoming 'the recognised name of a science.' Today, political economy, where it is not used as a synonym for economics, may refer to very different things, including Marxian analysis, applied public-choice approaches emanating from the Chicago school and the Virginia school, or simply the advice given by economists to the government or public on general economic policy or on specific proposals. A rapidly growing mainstream literature from the 1970s has expanded beyond the model of economic policy in which planners maximize utility of a representative individual toward examining how political forces affect the choice of economic policies, especially as to distributional conflicts and political institutions. It is now an area of study in universities.

5.1.1. Karl Marx

Karl Heinrich Marx (1818-1883) was born on May 5, 1818 in the city of Trier, Germany. His father was a lawyer who came from a long line of Rabbis, but had changed his faith to Protestantism in order to keep his job. Karl Marx went to the University of Bonn to study law when he was 17 years old. Here he became engaged to Jenny von Westphalen, whose father, Baron von Westphalen, influenced Marx to read Romantic literature and Saint-Simonian politics. Only a year later, Marx was moved by his father to the University of Berlin where he studied Hegelianism, influenced by Ludwig Feurbach and other Hegelians. He admired G.W.F. Hegel's dialectics and belief in historical inevitability, but Marx questioned the idealism and abstract thought of philosophy and maintained his belief that reality lies in the material base of economics. In distinct contrast to G.W.F. Hegel's concentration on the state in his philosophy of law, Marx saw civil society as the sphere to be studied in order to understand the historical development of humankind. In 1841 Marx earned his doctorate at Jena with his work on the materialism and atheism of Greek atomists. It was difficult for Marx to find publishers because of his radical political views, so he moved to Cologne, which was known to house a strong liberal opposition movement. The liberal group the Cologne Circle published a paper by Marx defending the freedom of the press in their newspaper The Rhenish Gazette (in 1842 he was made the editor of the paper). In Cologne Marx met Moses Hess, a radical who organized socialist meetings, which Marx attended. At these meetings Marx learned of the struggles of the German working-class. Based on the information he gathered from the members present at the meetings, Marx wrote an article on the poverty of the Mosel wine-farmers in which he was highly critical of the government. When the article was published in 1843, the Prussian authorities banned The Rhenish Gazette and threatened Marx with his arrest. Marx married his fiancé and they fled together to Paris. Here he took a position as editor of a political journal called *Deutsch- FranzösischeJahrbücher* (Franco-German Annals) that was designed to connect French socialism and adical Hegelianism. Although the journal only lived as long as one issue, it was a valuable opportunity for Marx. Through it he met his life-long friend Friedrich Engels, a contributor to the journal. Other prominent contributors included his old mentor from Berlin, Bruno Bauer, and the Russian anarchist Michael Bakunin. While in Paris, Marx became a communist, and worked primarily on studying political economy and the history of the French Revolution. He wrote a series of papers known as *–konomischphilosophischeManuskripteausdemJahre* (Economic and Philosophical Manuscripts, 1844), however they were not published until the 1930s. The Manuscripts are influenced by Feuerbach and outline a humanist idea of communism. Marx contrasts capitalist society, and an alienated nature of labour, with communist society, in which human beings in cooperative production develop their nature freely. In 1844 Marx reviewed Bruno Bauer's book On the Jewish Question. More than a review, Marx used the article to critique the continued influence of religion over politics, and propose a revolutionary change to the structure of European society.

In 1845 Marx was expelled from France by Guizot. He fled with Friedrich Engels to Brussels where they stayed for three years with intermittent trips to England to visit Engels' family who had cotton-spinning interests in Manchester. While in Brussels Marx wrote a piece against the idealistic socialism of P.J. Proudhon called The Poverty of Philosophy. He also worked on his materialist conception of history, and developed the manuscript that would come to be named The German Ideology when it was published after his death. This paper argues that the nature of an individual is dependent upon the material conditions that determine his production. It is a historical study of modes of production through the ages, and in it Marx predicts the collapse of industrial capitalism and the advancement of communism. Marx joined the Communist League at this time, which was an organization of German émigré workers centered in London. Marx and Engels became the major theoretical force of the League, and at a conference in 1847 they were commissioned to write a declaration of the League's position. The hope was that the *Manifest der kommunistischenPartei* (The Communist Manifesto) would inspire social revolution, and no sooner was it published than the 1848 revolutions broke out across Europe. This work marks a turn in Marx's writing from appealing to natural rights as justification for social reform, to indicating that the laws of history would inevitably lead to the power of the working class. The Manifesto distinguishes communism from other movements, proposes specific social reforms, and includes a description of the struggles between the proletariat and the bourgeoisie. It also explicitly encourages workers to unite in revolution against the existing regimes. The panic caused by the February revolution of 1848 caused the Belgian government to expel Marx from Brussels. He was invited by the French provisional government to return to Paris. From there, he returned to Cologne with some friends to start the newspaper the *NeueRheinischeZeitung*. The government there attempted to shut down the paper through legal means, and finally succeeded by finding pretexts to expel the editors. Marx and his friends were expelled after the revolts of May 1849, and the newspaper's last edition was June 1849. Marx had to return to Paris, but he was expelled again immediately, and moved on to London, which would be his final home.

In London Marx rejoined with the Communist League, confident that there would be further revolutionary action in Europe. He proceeded to write two pamphlets about the 1848 revolution in France and its effects, titled, The Class Struggles in France and The 18th Brumaire of Louis Bonaparte. He felt that new revolution would only be possible if there was to be a new crisis, and he hoped to uncover what would cause this crisis. He spent a large amount of his time in the British Museum studying political economy toward this end. For the first part of the 1850s Marx, Jenny, and their four children lived in an impoverished state in a three room flat in London's Soho. The couple would have two more children, but only three in all would survive. The family survived primarily on gifts from Friedrich Engels whose own income came from the family business in Manchester. Marx also earned a small amount from articles he wrote as the foreign correspondent for the New York Daily Tribune. In 1864 Marx and Friedrich Engels together founded the International Workingmen's Association, which would finally break up due to disagreements between Marx and the anarchist Mikhail Babuknin. By 1857 Marx had written an 800-page manuscript which was to become Das Kapital (Capital). This is his major work on political economy, capital, landed property, the state, wage labour, foreign trade and the world market. In the early part of the 1860s he took a break from his work on Das Kapital to work on Theories of Surplus Value, a three-volume work. This text discusses specific theories of political economy, primarily those of Adam Smith and David Ricardo. In 1867 Marx published volume I of Das Kapital, an analysis of the capitalist process of production, with an elaboration on his version of labour theory value, surplus value, and exploitation, that he predicted would lead to a falling profit rate and the collapse of industrial capitalism. Marx continued to work on Volumes II and III of Das Kapital for the rest of his life, even though they were essentially finished in the late 1860s. Friedrich Engels would publish the last two volumes after Marx's death. By 1871 Marx's daughter Eleanor, who was 17 at the time, was helping her father with his work. She had been taught at home by Marx himself, and grew up with a rich understanding of the capitalist system which would allow her to play an important part in the future of the British labour movement. Marx's health rapidly declined during the last ten years of his life and he was unable to work at the same impressive pace he had set in his early years. He still paid close attention to contemporary politics, especially concerning Germany and Russia, and he often offered his comments. In his Critique of the Gotha Programme he critiqued the actions of his admirers Karl Liebknecht and August Bebel, disagreeing with their compromises with state socialism in the interest of a united socialist party. He indicated in his letters to Vera Zasulich of this time that he imagined it could be possible for Russia to bypass a capitalist stage of development and move directly to communism by basing its economy on common ownership of land characterized by the village. In 1881 both Marx and his wife became ill. Marx had a swollen liver, and survived, but Jenny died on December 2, 1881. In January 1883 Marx was deeply saddened by the loss of his eldest daughter to cancer. On March 14, 1883 Marx passed away. He is buried at High gate Cemetery in London.

5.2. Economic Ideas of Marx

For Karl Marx, the basic determining factor of human history is economics. According to him, humans - even from their earliest beginnings - are not motivated by grand ideas but instead by material concerns, like the need to eat and survive. This is the basic premise of a materialist view of history. At the beginning, people worked together in unity and it wasn't so bad. But eventually, humans developed agriculture and the concept of private property. These two facts created a division of labour and a separation of classes based upon power and wealth. This, in turn, created the social conflict which drives society. All of this is made worse by capitalism which only increases the disparity between the wealthy classes and the labour classes. Confrontation between them is unavoidable because those classes are driven by historical forces beyond anyone's control. Capitalism also creates one new misery: exploitation of surplus value. For Marx, an ideal economic system would involve exchanges of equal value for equal value, where value is determined simply by the amount of work put into whatever is being produced. Capitalism interrupts this ideal by introducing a profit motive - a desire to produce an uneven exchange of lesser value for greater value. Profit is ultimately derived from the surplus value produced by workers in factories.

A labourer might produce enough value to feed his family in two hours of work, but he keeps at the job for a full day - in Marx's time, that might be 12 or 14 hours. Those extra hours represent the surplus value produced by the worker. The owner of the factory did nothing to earn this, but exploits it nevertheless and keeps the difference as profit. In this context, Communism thus has two goals: First it is supposed to explain these realities to people unaware of them; second it is supposed to call people in the labour classes to prepare for the confrontation and revolution. This emphasis on action rather than mere philosophical musings is a crucial point in Marx's program. As he wrote in his famous Theses on Feuerbach: 'the philosophers have only interpreted the world, in various ways; the point, however, is to change it.' But eventually, humans developed agriculture and the concept of private property. These two facts created a division of labour and a separation of classes based upon power and wealth. This, in turn, created the social conflict which drives society. All of this is made worse by capitalism which only increases the disparity between the wealthy classes and the labour classes. Confrontation between them is unavoidable because those classes are driven by historical forces beyond anyone's control. Capitalism also creates one new misery: exploitation of surplus value. For Marx, an ideal economic system would involve exchanges of equal value for equal value, where value is determined simply by the amount of work put into whatever is being produced.

Capitalism interrupts this ideal by introducing a profit motive – a desire to produce an uneven exchange of lesser value for greater value. Profit is ultimately derived from the surplus value produced by workers in factories. A labourer might produce enough value to feed his family in two hours of work, but he keeps at the job for a full day - in Marx's time, that might be 12 or 14 hours. Those extra hours represent the surplus value produced by the worker. The owner of the factory did nothing to earn this, but exploits it nevertheless and keeps the difference as profit. In this context, Communism thus has two goals: First it is supposed to explain these realities to people unaware of them; second it is supposed to call people in the labour classes to prepare for the confrontation and revolution. This emphasis on action rather than mere philosophical musings is a crucial point in Marx's program. As he wrote in his famous Theses on Feuerbach: 'the philosophers have only interpreted the world, in various ways; the point, however, is to change it.' Economics, then, are what constitute the base of all of human life and history – generating division of labour, class struggle, and all the social institutions which are supposed to maintain the status quo. Those social institutions are a superstructure built upon the base of economics, totally dependent upon material and economic realities but nothing else. All of the institutions which are prominent in our daily lives - marriage, church, government, arts, etc. - can only be truly understood when examined in relation to economic forces. Marx had a special word for all of the work that goes into developing those institutions: ideology. The people working in those systems - developing art, theology, philosophy, etc. - imagine that their ideas come from a desire to achieve truth or beauty, but that is not ultimately true. In reality, they are expressions of class interest and class conflict. They are reflections of an underlying need to maintain the status quo and preserve current economic realities. This isn't surprising - those in power have always wished to justify and maintain that power.

5.3. The Importance of the Study of Marxian Economics

In the context of present day economics many of the Marxian tenets such as his labour theory of value, the theory of price and wages, the theory of surplus value, the theory of crises, misery and capitalist break down, etc., are hardly valid. In fact the labour theory of value is inconsistent with the pricing process of the capitalist economy. Marx was not an advocate of socialism. Marx dealt mainly with capitalism and the process of its transformation. Yet surprisingly Marx is regarded as the prophet of Socialist faith. Hence we have to make a study of Marxian economics for the following reasons:

- A study of Marx will help us to understand the operation of capitalism, Marx was the most severe critic of capitalism. He revealed the Inner contradictions and weaknesses of capitalism. He also explained the fundamentals of social change. So by studying Marx, we can be aware of the shortcomings of capitalism and improve the system.
- Marxian economics teaches us that material well-being is the pre-requisite of social well-being. Arts, Science, Literature, Politics, Philosophy, in fact the entire human civilizations is highly influenced by economic factors. Marxism draws our attention to the miseries and privations of the common masses. It creates a consciousness in us about the urgency of rendering social justice.
- Marxian description of capitalism throws light on certain features of capitalism which were either not properly analyzed or not touched at all by the classical economists. Thus concentration of economic power, growth of monopoly and finance capital, unemployment, economic crisis due to overproduction and under consumption, imperialism, etc., have become highly popular topics of discussion today. But credit goes to Marx for drawing attention of his successors to these burning problems. He presented these in the form of an integrated analysis. It is thus imperative that we should have knowledge of- Marxian economics for a better comprehension of these issues.
- A study of Marxian economics is essential to acquaint ourselves with the Marxian terminology so as to understand modern socialism and interpret socialist writing in proper perspective. Socialist literature abounds in terms such as immiserisation, proletariat, organic composition of capital, dialectical materialism, etc. These are mostly of Marxian origin. Without an understanding of the implications of these Marxian phrases, our understanding of socialism would remain incomplete.
- Orthodox classical economists argued in terms of harmony of interests in the society. Butas Joan Robinson pointed out, Marxian conception of a permanent conflict of interests in economic life gives a more realistic picture of the Society. It is obvious that the interests of the capitalists and

workers are incompatible and hence would come into clashes. Hence a study of Marx would give us a true picture of na capitalist society.

- Economics is a Social Science. Economics studies the social relations of production and distribution. But in classical economics very often we find a divorce between economic theory and social relation Marxism was a reaction against this conception. In this context his Theo-of class struggle represented a more realistic approach, although the assumptions and conclusions were not always valid. Thus a study of Marxian economics enables us to understand the background of class struggle in the society.
- Marx's critique of Capitalism has not proved to be essential true. However, his analysis of capitalism as a passing stage in the historical evolution of Society provides interesting reading.
- A study of Marx enables us to understand the nature and role the state. In a capitalist economy the state is an instrument of oppressor. In the ultimate communist society, the state will wither away. But there scope to manipulate the state to further the interests of the masses.
- Knowledge of Marxian ideas is highly essential to understand the contemporary economic society. Marxism today is the exclusive method of social analysis for a third of the human race. So it would absurd to comprehend contemporary social reality without comprehending Marxism.

5.3.1. Marxian Poliitical Economy: Stages of Development

According to Marx, human civilization has manifested itself in a series of organizational structures, each determined by its primary mode of production, particularly the division of labour that dominates in each stage.

- ***The tribal form.*** Tribal society has no social classes but is structured around kinship relations, with hunting the province of men and domestic work the province of women. The tribal form, according to Marx and Engels, is quite elementary at this stage, 'a further extension of the natural division of labour existing in the family'. During this stage, it is also possible to see a slave culture established, particularly as the population increases, leading to 'the growth of wants' and the growth of relations with outside civilizations (through war or barter). With slave culture, we see the beginning of class society.
- ***Primitive communism***: 'the ancient communal and State ownership which proceeds especially from the union of several tribes into a city by agreement or by conquest'. During this stage, the concept of private property begins to develop: 'With the development of private property, we find here for the first time the same conditions which we shall find again, only on a more extensive scale, with modern private property. On the one hand, the concentration of private property.; on the other hand, coupled with this, the transformation of the plebeian small peasantry into a proletariat'.

- ☆ ***Feudal or estate property***: 'Like tribal and communal ownership, it is based again on a community; but the directly producing class standing over against it is not, as in the case of the ancient community, the slaves, but the ensured small peasantry'. In the city, the feudal structure manifested itself in trade guilds. The organization of both the country and the city 'was determined by the restricted conditions of production - the small-scale and primitive cultivation of the land, and the craft type of industry', which meant that there 'was little division of labour in the heyday of feudalism'. Exploitation functioned differently during stage than during the height of capitalism because each feudal peasant knew exactly what proportion of his labour had to be handed over to the aristocracy and the church; the rest was his or hers to use.
- ☆ ***Capitalism***: Because of the eventual growth of commerce (and of human populations), feudal society began to accumulate capital, which, along with the increased debt incurred by the aristocracy, eventually led to the English Revolution of 1640 and the French Revolution of 1789, both of which opened the way for the establishment of a society structured around commodities and profit (*i.e.* capitalism). In such a society, the proletariat is fooled into believing that s/he is free because s/he is paid for his/her labour. In fact, the transformation of labour into an abstract quantity that can be bought and sold on the market leads to the exploitation of the proletariat, benefitting a small percentage of the population in control of capital. The working class thus experiences alienation since the members of this class feel they are not in control of the forces driving them into a given job. The reason for this situation is that someone else owns the means of production, which are treated like private property.

Marx himself considered his theory of surplus-value his most important contribution to the progress of economic analysis (Marx, letter to Engels of 24 August 1867). It is through this theory that the wide scope of his sociological and historical thought enables him simultaneously to place the capitalist mode of production in his historical context, and to find the root of its inner economic contradictions and its laws of motion in the specific relations of production on which it is based. The price of commodity produced is determined by the labour involved in that commodity. In the complex capitalist environment labourer provides his services to his boss in order to produce a commodity but in response to it he just gets a small chunk of the profit. The remaining profit goes to the boss or who is responsible to conducting that business. The theory of surplus value says that this labourer, his efforts are helpful, meaningful to the owner of that business as the surplus amount goes to the boss instead of that labourer who is actually responsible for carrying out the whole business. Thus, Karl Marx is of the view that a labourer and the person who is carrying out the business be treated on the equal grounds but the capitalist economy does not take care of this fact.

Karl Marx criticized this concept of dividing the class between the 'haves' and 'have-nots'. In the Marx theory of class conflict he enunciated the impacts of theory of surplus value. Surplus value generates a handsome income for the 'owner' however it gives the diminishing returns to the labour. Hence, this gives birth to petit bourgeois and proletariat. The former are those who are rich, wealthy and influential in the society and the latter are those who are weak, poor and remain at the discretion of the influential persons. Therefore, surplus value generates a class conflict between the different strata of the people. Secondly, surplus value gives birth to alienation. This means that a person who is capable and talented yet, he cannot impart his full abilities to the work, led him to alienation. The reason behind is that the boss is just concerned about his product regardless of the talent in the labour and on the other hand, the labour being at the mercy of poor environmental conditions, he is supposed to do the work under the wage system. The poor labourer does not have any other option to make his livelihood. Thirdly, the theory of alienation, class conflict and surplus value give birth to the polarization in the society. The poor are getting poorer and the rich are getting richer. Thus, such kind of exploitation paralyses the society. So polarization creates a sense of deprivation amongst the members of the society. Fourthly, according to Marxist, surplus value gives birth to wars and arm conflict. The reason is that the additional amount or profit earned by the big cartels is utilized on the wars. Thus, gives birth to chaos and anarchy. Therefore, the Marxist school of thought believes in the closed economy.

Karl Marx accepted Ricardo's labour theory of value (that the value of a product is based on the quantity of labour that went into producing it), but he subscribed to a subsistence theory of wages for a different reason than that given by the classical economists. In Marx's estimation, it was not the pressure of population that drove wages to the subsistence level but rather the existence of large numbers of unemployed workers. Marx blamed unemployment on capitalists. He renewed Ricardo's belief that the exchange value of any product was determined by the hours of labour necessary to create it. Furthermore, Marx held that, in capitalism, labour was merely a commodity: in exchange for work, a labourer would receive a subsistence wage. Marx speculated, however, that the owner of capital could force the worker to spend more time on the job than was necessary for earning this subsistence income, and the excess product–or surplus value–thus created would be claimed by the owner. This argument was eventually disproved, and the labour theory of value and the subsistence theory of wages were also found to be invalid. Without them, the surplus-value theory collapsed.

As said before, Marx's theory of classes is based on the recognition that in each class society, part of society (the ruling class) appropriates the social surplus product. But that surplus product can take three essentially different forms (or a combination of them). It can take the form of straightforward unpaid surplus labour, as in the slave mode of production, early feudalism or some sectors of the Asiatic mode of production (unpaid corvée labour for the Empire). It can take the form of goods appropriated by the ruling class in the form of use-values pure and simple (the products of surplus labour), as under feudalism when feudal rent is paid in a certain amount of produce (produce rent) or in its more modern remnants, such as sharecropping. And it can take a money form, like money-rent in the final phases of feudalism, and capitalist profits. Surplus-value is essentially just that: the money form of the social surplus product or, what amounts to the same, the money product of surplus labour. It has therefore a common root with all other forms of surplus product: unpaid labour.

This means that Marx's theory of surplus-value is basically a deduction (or residual) theory of the ruling classes' income. The whole social product (the net national income) is produced in the course of the process of production, exactly as the whole crop is harvested by the peasants. What happens on the market (or through appropriation of the produce) is a distribution (or redistribution) of what already has been created. The surplus product, and therefore also its money form, surplus-value, is the residual of that new (net) social product (income) which remains after the producing classes have received their compensation (under capitalism: their wages). This 'deduction' theory of the ruling classes' income is thus ipso factor an exploitation theory. Not in the ethical sense of the word - although Marx and Engels obviously manifested a lot of understandable moral indignation at the fate of all the exploited throughout history, and especially at the fate of the modern proletariat - but in the economic one. The income of the ruling classes can always be reduced in the final analysis to the product of unpaid labour: that is the heart of Marx's theory of exploitation. That is also the reason why Marx attached so much importance to treating surplus-value as a general category, over and above profits (themselves subdivided into industrial profits, bank profits, commercial profits etc.), interest and rent, which are all part of the total surplus product produced by wage labour. It is this general category which explains both the existence (the common interest) of the ruling class (all those who live off surplus value), and the origins of the class struggle under capitalism.

Marx likewise laid bare the economic mechanism through which surplus-value originates. At the basis of that economic mechanism is a huge social upheaval which started in Western Europe in the 15th century and slowly spread over the rest of the continent and all other continents (in many so-called underdeveloped countries, it is still going on to this day). Through many concomitant economic (including technical), social, political and cultural transformations, the mass of the direct producers, essentially peasants and handicraftsmen, are separated from their means of production and cut off from free access to the land. They are therefore unable to produce their livelihood on their own account. In order to keep themselves and their families alive, they have to hire out their arms, their muscles and their brains, to the owners of the means of production (including land). If and when these owners have enough money capital at their disposal to buy raw materials and pay wages, they can start to organise production on a capitalist basis, using wage labour to transform the raw materials which they buy, with the tools they own, into finished products which they then automatically own too. The capitalist mode of production thus presupposes that the producers' labour power has become a commodity. Like all other commodities, the commodity labour power has an exchange value and a use value. The exchange value of labour power, like the exchange value of all other commodities, is the amount of socially necessary labour embodied in it, *i.e.* its reproduction costs. This means concretely the value of all the consumer goods and services necessary for a labourer to work day after day, week after week, month after month, at approximately the same level of intensity, and for the members of the labouring classes to remain approximately stable in number and skill (*i.e.* for a certain number of working-class children to be fed, kept and schooled, so as to replace their parents when they are unable to work anymore, or die). But the use value of the commodity labour power is precisely its capacity to create new value, including its potential to create more value than its own reproduction costs. Surplus value is but that difference between the total new value created by the commodity labour power, and its own value, its own reproduction costs. The whole Marxian theory of surplus-value is therefore based upon that subtle distinction between 'labour power' and 'labour' (or value). But there is nothing 'metaphysical' about this distinction. It is simply an explanation (demystification) of a process which occurs daily in millions of cases.

The capitalist does not buy the worker's 'labour'. If he did that there would be obvious theft, for the worker's wage is obviously smaller than the total value he adds to that of the raw materials in the course of the process of production. No: the capitalist buys 'labour power', and often (not always of course) he buys it at its *justumpretium*, at its real value. So he feels unjustly accused when he is said to have caused a 'dishonest' operation. The worker is victim not of vulgar theft but of a social set-up which condemns him first to transform his productive capacity into a commodity, then to sell that labour power on a specific market (the labour market) characterised by institutional inequality, and finally to content himself with the market price he can get for that commodity, irrespective of whether the new value he creates during the process of production exceeds that market price (his wage) by a small amount, a large amount, or an enormous amount. The labour power the capitalist has bought 'adds value' to that of the used-up raw materials and tools (machinery, buildings etc.). If, and until that point of time, this added value is inferior or equal to the workers' wages, surplus-value cannot originate. But in that case, the capitalist has obviously no interest in hiring wage labour. He only hires it because that wage labour has the quality (the use value) to add to the raw materials' value more than its own value (*i.e.* its own wages). This 'additional added value' (the difference between total 'value added' and wages) is precisely surplus-value. Its emergence from the process of production is the precondition for the capitalists' hiring workers, for the existence of the capitalist mode of production.

The institutional inequality existing on the labour market arises from the very fact that the capitalist mode of production is based upon generalised commodity production, generalised market economy. This implies that a property less labourer, who owns no capital, who has no reserves of larger sums of money but who has to buy his food and clothes, pay his rent and even elementary public transportation for journeying between home and workplace, in a continuous way in exchange of money, is under the economic compulsion to sell the only commodity he possesses, to wit his labour power, also on a continuous basis. He cannot withdraw from the labour market until the wages go up. He cannot wait. But the capitalist, who has money reserves, can temporarily withdraw from the labour market. He can lay his workers off, can even close or sell his enterprise and wait a couple of years before starting again in business. The institutional difference makes price determination of the labour market a game with loaded dice, heavily biased against the working class. One just has to imagine a social set-up in which each citizen would be guaranteed an annual minimum income by the community, irrespective of whether he is employed or not, to understand that 'wage determination' under these circumstances would be quite different from what it is under capitalism. In such a set-up the individual would really have the economic choice whether to sell his labour power to another person (or a firm) or not. Under capitalism, he has no choice. His is forced by economic compulsion to go through that sale, practically at any price. The economic function and importance of trade unions for the wage-earners also clearly arises from that elementary analysis. For it is precisely the workers'' combination' and their assembling a collective resistance fund (what was called by the first French unions *caisses de résistance*, 'reserve deposits') which enables them, for example through a strike, to withdraw the supply of labour power temporarily from the market so as to stop a downward trend of wages or induce a wage increase. There is nothing 'unjust' in such a temporary withdrawal of the supply of labour power, as there are constant withdrawals of demand for labour power by the capitalists, sometimes on a huge scale never equaled by strikes. Through the functioning of strong labour unions, the working class tries to correct, albeit partially and modestly, the institutional inequality on the labour market of which it is a victim, without ever being able to neutralize it durably or completely.

It cannot neutralize it durably because in the very way in which capitalism functions there is a powerful built-in corrective in favour of capital: the inevitable emergence of an industrial reserve army of labour. There are three key sources for that reserve army: the mass of pre-capitalist producers and self-employed (independent peasants, handicraftsmen, trades-people, professional people, small and medium-sized capitalists); the mass of housewives (and to a lesser extent, children); the mass of the wage-earners themselves, who potentially can be thrown out of employment. The first two sources have to be visualised not only in each capitalist country seen separately but on a world scale, through the operations of international migration. They are still unlimited to a great extent, although the number of wage-earners the world over (including agricultural wage labourers) has already passed the one billion mark. As the third source, while it is obviously not unlimited (if wage labour would disappear altogether, if all wage labourers would be fired, surplus-value production would disappear too; that is why 'total robotism' is impossible under capitalism), its reserves are enormous, precisely in tandem with the enormous growth of the absolute number of wage earners.

The fluctuations of the industrial reserve army are determined both by the business cycle and by long-term trends of capital accumulation. Rapidly increasing capital accumulation attracts wage labour on a massive scale, including through international migration. Likewise, deceleration, stagnation or even decline of capital accumulation inflates the reserve army of labour. There is thus an upper limit to wage increases, when profits (realised profits and expected profits) are 'excessively' reduced in the eyes of the capitalists, which triggers off such decelerated, stagnating or declining capital accumulation, thereby decreasing employment and wages, till a 'reasonable' level of profits is restored. This process does not correspond to any 'natural economic law' (or necessity), nor does it correspond to any 'immanent justice'. It just expresses the inner logic of the capitalist mode of production, which is geared to profit. Other forms of economic organization could function, have functioned and are functioning on the basis of other logics, which do not lead to periodic massive unemployment. On the contrary, a socialist would say - and Marx certainly thought so - that the capitalist system is an 'unjust', or better stated 'alienating', 'inhuman' social system, precisely because it cannot function without periodically reducing employment and the satisfaction of elementary needs for tens of millions of human beings.

Marx's theory of surplus-value is therefore closely intertwined with a theory of wages which is far away from Malthus's, Ricardo's or the early socialists' (like Ferdinand Lassalle's) 'iron law of wages', in which wages tend to fluctuate around the physiological minimum. That crude theory of 'absolute pauperization' of the working class under capitalism, attributed to Marx by many authors, is not Marx's at all, as many contemporary authors have convincingly demonstrated. Such an 'iron law of wages' is essentially a demographic one, in which birth rates and the frequency of marriages determine the fluctuation of employment and unemployment and thereby the level of wages. The logical and empirical inconsistencies of such a theory are obvious. Let it be sufficient to point out that while fluctuations in the supply of wage-labourers are considered essential, fluctuations in the demand for labour power are left out of the analysis. It is certainly a paradox that the staunch opponent of capitalism, Karl Marx, pointed out as early as in the middle of the 19th century the potential for wage increases under capitalism, even though not unlimited in time and space. Marx also stressed the fact that for each capitalist, wage increases of other capitalists' workers are considered increases of potential purchasing power, not increases in costs.Marx distinguishes two parts in the workers' wage, two elements of reproduction costs of the commodity labour power. One is purely physiological, and can be expressed in calories and energy quanta; this is the bottom below which the wage cannot fall without destroying slowly rapidly the workers' labour capacity. The second one is historical-moral, as Marx calls it, and consists of those additional goods and services which a shift in the class relationship of forces, such as a victorious class struggle, enables the working class to incorporate into the average wage, the socially necessary (recognized) reproduction costs of the commodity labour power (*e.g.* holidays after the French general strike of June 1936). This part of the wage is essentially flexible. It will differ from country to country, continent to continent and from epoch to epoch, according to many variables. But it has the upper limit indicated above: the ceiling from which profits threaten to disappear, or to become insufficient in the eyes of the capitalists, who then go on an 'investment strike'. So Marx's theory of wages is essentially an accumulation-of-capital theory of wages which sends us back to what Marx considered the first 'law of motion' of the capitalist mode of production: the compulsion for the capitalists to step up constantly the rate of capital accumulation.

5.3.2. Surplus Value

Surplus value is the difference between the value of what an employee produces and what the employee is paid – the surplus value is converted into the owner's profit. This is a complicated concept and initially seems counter-intuitive. Machinery is a part of modern production and does not machinery increase efficiency? The machine presumably costs less over its life than the worker; isn't that why capitalists buy machines, so they can employ fewer workers and increase productivity? True on both counts. But the value of machines is consumed in production - their value is transferred to the products that are produced with them. It is the physical labour of production that produces the commodity that is sold for more than was paid for the materials used to make it. This concept is easier to understand when it is applied across the life of a commodity rather than narrowly within only the enterprise that manufactures the final product. Any product made for sale has an 'exchange value.' This value is not necessarily the same as its 'use value' - the intrinsic value a product has to the user of it. If it takes eight hours for an individual to make a shirt for herself, then the shirt might be said to have a use value of eight hours of labour. Perhaps instead of wearing it herself the shirt maker wishes to barter the shirt for a pair of shoes. If the shoes require sixteen hours to make, the shoe maker is not likely to see that as a fair exchange. But if the shoe maker needs two shirts, then the labour that went into each side of the exchange is equal (assuming the skill and intensity of work are close to equal). In this example, the pair of shoes can be said to have the value of two shirts.

In a modern capitalist economy, the shirt or shoe is sold for money - its exchange value is the amount of money paid for it. But the shirt maker working for a wage paid by a manufacturer will receive only a portion of that value - the difference, the surplus value, is the source of profit. If the capitalist willingly paid to his employees the full value of what they produce, he wouldn't be a capitalist - there would be no profit. The owner of the factory is not altruistic - he intends to extract surplus value. But that owner does not keep all the surplus value - he must share it with those who help circulate the commodity. Distributors and merchants assume the cost of circulation, part of the expense of a commodity, while sharing the surplus value. The distributor has specialized skills and can circulate the commodity more efficiently than the manufacturer; because the cost of circulating the commodity is thereby reduced, there is more surplus value to be shared. In the following hypothetical case, the surplus value is shared with the distributor and the merchant. Let's say the factory owner pays a wage that is equal to eight dollars to each worker for each widget. The owner sells the widget to the distributor for ten dollars, the distributor sells it to the merchant for twelve dollars and the merchant sells the widget for fourteen dollars. When the worker goes to the store to buy a widget, she pays fourteen dollars although she was paid only eight dollars to make one. Thus, the widget is worth six dollars more than what the factory owner paid to the worker, not the two-dollar difference between the wage and what the factory owner received for it. The distribution of that surplus value can change among the capitalists. These capitalists compete against each other to earn a bigger profit, at the same time they cooperate in getting the product to market. The widget manufacturer might miscalculate the demand and overproduce, causing a glut that reduces the price that can be realized. Or a giant merchant chain becomes so big that it has the power to force lower prices – the chain wishes to sell the widget for less to undercut its smaller competitors, and possesses sufficient clout by virtue of its size to negotiate a discount, forcing the manufacturer to cut its wholesale prices.

If the manufacturer does not wish to see its profits reduced, it has to reduce its costs. The primary way it can do so is to lower its labour-power costs. This can mean cuts to wages or benefits, increased workloads, layoffs or moving production somewhere else. In each of these cases, the capitalist is buoying profit levels by extracting more surplus value. More will be extracted from the workforce through suppressing pay or an intensification of work. The above example is of course an oversimplification. The factory owner has costs other than labour power, and employees do not create the widgets solely with bare hands. (And, in reality, the employee will be paid far less than the 80 percent of the factory owner's proceeds in our hypothetical example.) There is machinery in the manufacturing process, and raw materials (including previously manufactured components) are needed to make the widgets. If the company's shares are traded on a stock exchange, the shareholders will be expecting a hefty cut of the profits. Labour power is the source of surplus value because raw materials and the value of the machinery are consumed in production while labour power produces more value than is paid for it. That does not mean that machines aren't productive nor that they don't raise the productivity of those who work with them. They do both. The surplus value contained in the machines placed in production was realized by the manufacturer of the machine upon selling it; the machines transfer their value to the commodities produced using them. (Payments might continue to be made on the machine after it is put into service, but the payments go to the lender who financed the machine's purchase; interest is another sharing of surplus value. Paying rent is as well.) A commodity is produced with direct labour, machines and raw materials, but the machines and raw materials assist labour in producing the surplus value - machines make labour more productive, enabling more surplus value to be extracted from each employee. (One worker using a bulldozer can do as much as several workers with shovels.Computerization also reduce the number of employees in an office; more work is done with fewer people.) Raw materials and other commodities are bought by the capitalist so they can be sold in a new form for a higher price. Raw materials and natural resources can't do that by themselves -labour power is the only commodity that can add the value that becomes surplus value.

5.4. Marxian Political Economy: Theory of Capitalist Crisis

Marx's materialistic interpretation of history is also known as economic interpretation of history. It implies that economic factors play the decisive role in shaping the evolution of society. It believes that 'the ultimate determinant of social change is not to be found in his (man's) ideas of external truth and social justice but in changes in the mode of production and exchange. According to this theory at any given time, people in a society possess a certain level of productive ability. It depends on tools, machines raw materials, natural resources, etc. It also depends on the people themselves-on their knowledge, talent, and efficiency, etc. Together these are known as 'Productive forces'. According to Marx, these productive forces determine the way people make their living. To earn a living people have to engage themselves in some form of economic activity. People have numerous wants. Satisfaction of these wants necessitates productive activity. Whenever men cooperate together to produce and exchange the means of satisfying their economic wants they enter into certain productive relations with each other. These relations are indispensable and independent of their will. Production involves a relation between man and man. At any particular time these productive relationships correspond to a definite stage of development of the forces of production. The productive forces determine the relations of production. For example, under slavery people are related to each other in producing and exchanging the means of life as master and slave, under feudalism as lord and serf, under capitalism as capitalist and worker. These production and exchange relationships have' been called by Marx as 'relations of production.'

In an agricultural society, for example, land would be considered as the principal form of wealth: The ownership of land would provide the clue to its superstructure of social, political, and cultural institutions, and ideas. The landowners will constitute the ruling class. They will receive the highest respect. Similarly, under capitalism the capitalists constitute the predominant class. The state is dominated by the capitalists. Capitalist property ownership influences the legal, social and cultural institutions and mental conceptions. The sum total of the relations of production in all fields of productive activity together constitutes the 'economic structure of society'. The economic society moulds the superstructure of society. The superstructure contains the human ideas, social, political, legal institutions, intellectual activities, family structure, art forms and spiritual processes. It thus contains the institutions and activities that support the class structure of society. As the economic structure changes, elements of the superstructure also change. With the passage of time productive forces develop, Sooner or later a time conies when the developing productive forces come in conflict with the existing class structure that is the social relations of production. Further improvement of the productive forces is impeded by the prevailing relations of production or the system of property. These relations which once promoted development in the methods of production, no stand in the way of that development. There is a conflict between the sociocultural environment and the further development of the productive powers of the economy. The growing contradiction or incompatibility gives rise to a class struggle, which ensues between the rising class associated with the new means of production and the old ruling class whose influence is declining. Gradually the contradiction intensifies until as a result of a social revolution new relations of production compatible with the superior productive forces are established. In the middle ages, the merchants and craftsmen challenged the feudal lord and wrested the economic power from him. Under capitalism the wage earners will challenges the capitalist and do the same. 'The history of all hitherto existing societies is the history of class struggle.'

Marx's theory of social development is also known as 'dialectical materialism.' Marxian 'materialism implies that ideas, institutions, philosophies, religions, etc., all are determined by real material condition of life. This materialism is also at the root of Marxian concept of social classes. Marxian materialism simply means that men's economic activities are fundamental and they determine the general way the character of everything else they do. Marx got the idea of 'dialectics' from the German philosopher Hegel. According to the dialectical principle everything includes it's opposite or negative. A thesis has an antithesis. The conflict of these opposites leads to the transformation of this thesis into a synthesis. In other words, the basic cause of change in all things in their self-movement lies in the struggle of opposites or in their internal contradiction. The contradiction gives rise to a synthesis between the thesis and antithesis. Human societies proceed from stage to stage of their development so that within each stage fundamental contradictions arise. When these contradictions are resolved, the society passes from one stage to another.

Marx's future-oriented perspective has its basis in his materialist conception of history. He suggests that the ways societies provide for their material well-being affects the type of relations that people will have with one another, their social institutions, and the prevailing ideas of the day. Marx uses the term 'the forces of production' to refer to the ways in which people provide for their needs. He uses the term 'relations of production' to describe social relationships that dominate the productive capacities of a society. Under capitalism, the forces of production lead to a set of relations of production which pit the capitalist and the proletariat against one another. To change the relations of production, Marx felt revolution was necessary. Revolution arises from exploited classes agitating for change in the relations of production that favour transformations in the forces of production. Marx applied the doctrine of dialectical materialism to analyze the historical evolution of society. Transition from one stage of social development to another is not caused by new principles or new ideas of truth and justice. What actually happens is that, in course of time, as productive forces change, they come into conflict with the existing relations of production. Inherent contradictions develop. The relations of production instead of furthering progress rather hinder progress. A social revolution takes place. Owing to a change in the economic system of society the superstructure is thrown aside. The transition thus takes place through a struggle between the classes. The victor in each becomes the vanquished in the next.

Marx has broadly distinguished four different stages through which the society has passed. These are primitive communism, slavery, feudalism and capitalism. The stage of primitive communism is characterized by a classless society. The means of production are socially owned. If there was any division of labour, it was limited to family. In course of time pastoral habits developed. People learnt how to tame animals. People learnt weaving and rudimentary forms of mining developed. All these led to the development of exchange. With the change in the forces of production, there was a great increase in the demand for labour, to carry on the new productive activities. This could not be satisfied under the existing form of economic society. The widening field of production brought slavery into existence. In other words, primitive communism was changed into slavery. Under slavery a new relation of production is established. The slave owner owned the means of production. There was further division of labour. Exchange flourished. New Institutions in the form of private ownership of land, money, interest, etc., developed. The society became more complex. States emerged to bring about coordination among the divergent interests. In course of time productive forces further developed. Then slavery gave way to feudalism. As we have noted elsewhere under feudalism, the feudal lord partially owned the means of production. The serfs or tenants were tied to the land, but they also owned some property. In the countryside agriculture flourished. In the urban areas small manufacturing handicraft used to be privately owned and operated by craft men. Production and exchange expanded. The scale of production under feudalism was very small and limited, which came in conflict with the growing productive powers which were slowly emerging. Besides the feudal, social and political forms also proved to be a hindrance to the further growth of the production powers of the economy. For instance, in England, since the 16th Century the bourgeoisie gradually became stronger on the basis of ne productive forces, such as new machines, factory processes, technical knowledge, superior weaponry, etc. These productive forces enabled them to challenge the supremacy of older feudal classes whose power and position depended on land rights, control of rural work forces and the prevailing social and legal systems. Conflicts followed. There was a series of revolutions. Ultimately, the economic structure changed from feudalism to capitalism. There was a transformation of the class structure from one consisting of lords and serfs to one of capitalists and workers. Under the capitalist mode of production, the capitalist owns the necessary means of production.

The worker owns none. The worker owns none. He is forced to work for the capitalist. From small-scale production there is a change to large-scale production. Production is for the market to earn more and more profit. Division of labour becomes more and more complex. There is free competition and freed of enterprise. Capitalism also is not free from this process of change. Growth of industrial capitalism introduced machines on a vast scale. This led to large-scale production, centralization, monopoly, finance capital and colonialism. In due process this system also developed its internal contradictions. Side by side, the capitalist mode of production will create a class of exploited men and then, unite, and strengthen this proletariat class which will overthrow capitalism. Under the fifth stage of socialism, which has not arrived yet, the workers themselves will own the means of production. The inherent contradictions within capitalism will be eliminated. Production will reach its fullest development. In this process of historical development of mankind, each successive stage represents an advance upon the preceding stage. Once a system outlives its utility and becomes an obstacle to the further development of the, productive forces, it will disintegrate, in the midst of a social revolution. This social revolution does not take place due to any religious, political or ideological reason as people used to believe before Marx. Marx held that the fundamental cause that governed every great movement in history was an economic cause. And historical evolution took place in accordance with the dialectic principle.

Marx was of the view that capitalism was destined to face disintegration ultimately. Marx believed that capitalism contained the seeds of its own destruction. He held that capitalism will disintegrate because of its inherent conflicts and contradictions. Following are the inherent conflicts and contradictions:

A. Alienation (Marx's Moral Critique of Capitalism)

Under capitalism, the relationship between labour and human expression changes: rather than labouring to fulfill their needs or express ideas, workers do so at the demands of capitalists. Workers are alienated from their labour because it no longer belongs to the worker, but rather to the capitalist. This alienates workers in four ways:

- ☆ Workers are alienated from their productive activity, in that they no longer labour to satisfy their own needs.
- ☆ Workers are alienated from the product of their labour, which now belongs to the capitalist. Instead of finding expression in producing, workers turn to consuming to express themselves.
- ☆ The cooperative nature of work is destroyed through the organization of the labour process, alienating workers from their fellow workers. Additionally, workers often must compete against one another for work and pay.
- ☆ Workers are alienated from their human potential, as the transformative potential of labour is lost under capitalism.

B. Exploitation

Exploitation is a set of social relations on which capitalism is built. Capitalists exploit workers by paying them less in wages than the value they produce. While a worker may earn eight dollars a day in wages, s/he may produce ten dollars a day worth of value, creating what Marx called surplus value. Capital grows by exploiting workers to generate ever greater amounts of surplus value, usually by lowering workers' wages. In addition, capitalists constantly compete with one another over capital by finding new ways to generate profit and surplus value in order to maintain an edge. Marx calls this drive the general law of capitalist accumulation. Capitalism is not the only historical epoch in which individuals are exploited, but it is the only one in which the mechanisms of exploitation are hidden behind independent, objectified, and reified structures, such as the market.

C. Falling Tendency of Rate of Profit

The main objective of the capitalists is to make profit. But in a capitalist society as capital accumulates, more proportion of it will consist of constant capital. Constant capital is the product of past labour and it does not create any new value. Only variable capital can create value. Although constant capital does not produce any surplus value yet the use of constant capital like machinery etc. increases the productivity of labour. So by using constant capital, the capitalist can reduce his cost, undersell his competitors and gain a temporary advantage over them. Competition therefore compels the capitalists' continually to increase their constant capital. Now, as the rate of profit is the ratio between surplus value and constant plus variable capital, the average rate of profit inevitably declines. Thus, in the short run it may be possible to offset the tendency of the rate of profit to fall by increasing the productivity of' labour. But in the long run the increasing organic composition of capital will bring about a sure decline in the rate of profit. The rate of profit is an indicator of the economic health of market capitalism. A declining rate of profit indicates that capitalism is in the process of decomposition.

D. Reserve Army of Labour and Unemployment

Marx assumed that under capitalism there was a tendency to produce a surplus labour force. He called it as the 'reserve army of the unemployment.' In the normal process of capitalist accumulation and production, in a mature capitalist economy, there is a tendency for a progressive decline in the amount of variable capital. This will lead to a decline in employment opportunities. With the growing use of machinery the demand for labour in the industries would progressively fall. Machines would increasingly devour men. The larger the quantity of capital per worker, the greater the productivity of labour, the smaller the number of workers employed to produce a certain quantity of output. With the increase in the size of individual enterprises consequent upon concentration and centralization of capital, the proportion of constant to variable capital would increase. The recurring economic crises also will vastly add to the army of unemployed. All these factors would tend to create an industrial reserve army or relative surplus population under capitalism.

E. The Increasing Misery of the Proletariat

Marx further held that with the progress of capitalism there was a tendency towards pauperisation or immiserisation of labour. His doctrine of increasing misery of labour has been interpreted in various ways. First, the condition of the -proletariat worsened due to unemployment. Due to the presence of a vast 'reserve army' the wages of the workers would be maintained at the minimum level. Intense competition among the workers would reduce the price of labour. Competition would also lead to over-work which contributed to their high mortality rate. Thus sometimes there is an absolute material deterioration. The economic success of capitalism followed by high production and wealth intensifies proletarian misery. The capitalist tries to realize the maximum possible surplus value from the worker. As the rate of profit of the capitalist falls, he becomes ever more desperate to raise the rate of exploitation. He lengthens the working day to the maximum extent possible. It even 'usurps the time for growth, development and healthy maintenance of the body' of the worker. The worker's wife children are thrown 'beneath the wheels of the juggernaut of capital.' There is mental degradation, anguish and growth of insecurity. Thus the downtrodden workers will grow increasingly miserable from the economic point of view.

F. The Increasing Severity of Economic Crises

The internal contradictions of the capitalist system manifest themselves in periodic economic crises. As Capitalism grows, economic crises take place with increasing severity. However Marx did not develop any separate business cycle theory in a systematic manner. However from his overall business cycle analysis, we may deduce three distinct theories of capitalist crisis. These are, (i) crisis associated with the falling rate of profit, (ii) crisis arising from disproportionality and (iii) under-consumption theory of crisis. First, in the process of capitalist development there in a relative decrease of variable as compared to constant capital and total capital. There is a rising organic consumption of capital. The degree of labour exploitation remaining the same, the rate of surplus value expresses itself in a continuously falling average rate of profit. Under capitalism production is guided by profit motive. With falling tendency in the rate of profit further accumulation is affected. It also poses as a threat to the production process. It also promote overproduction, speculation, crisis and surplus capital along with surplus population.

Secondly, according to Marx, crises under capitalism occur from disproportionality. This disproportionality takes place due to two basic characteristics of the capitalist system. First, there is 'anarchy of production'. It means that m a capitalistic economy productive decisions are made by. A large number of independent producers. Each producer has to anticipate the market demand for him. He has no knowledge about the moves of his rivals. The only factor to guide him is the profit motive. Secondly, there is no systematic plan or orderliness in capitalist production. There is absolutely no understanding or coordination among the various producers. Each stands for himself. The inevitable consequence is the bewildering lack of coordination among the different branches of production. Consequently, periodic disequilibrium arises between demand and supply. Economic crisis is thus inevitable in a capitalist economy and nothing in particular can be done to stop it. If we sincerely try to remedy the situation, capitalism would not be capitalism. To abolish crisis, capitalism must be abolished.

Thirdly, Marx also held that crises were caused by under consumption which had its roots in the inner contradictions of capitalism. The capitalist economy generates enormous productive powers. In fact, it has a tendency to develop the productive power on an ever-expanding scale. With this there is large-scale production of consumption goods, intended for sale. But under the prevailing pattern of distribution of income, the consuming power of the society in limited. Here by 'consuming power' we mean effective demand which depends on income. They receive subsistence wage. Unemployment further aggravates their purchasing power. The capitalistic distribution of income is in conflict with capitalistic production. The consumption power of the society is further restricted by the tendency of the capitalists to accumulate, so the demand for consumption goods cannot keep pace with the supply. In other words, there is an increasing gap between aggregate supplies on the one hand and consumption demand on the other. Economic crises naturally follow.

Another prediction of Marx regarding the future of capitalism is the concentration and centralization of capital. Marx's theory of concentration speaks of the tendency of the capitalist process 'to increase the size both of industrial plants and of units of control.' The concentration of capital implies the tendency for capital as a whole to accumulate. The capitalist has an endless passion for accumulation. The capitalists converted a large part of their surplus value into capital, In turn more capital meant the ability to enjoy more workers, more exploitation, more surplus value, and more capital and so on and on. Capital thus tended to increase in total size like a snowball rolling downhil1. According to Marx, under capitalism this process of capital accumulation would continue far into the future. Another development which would take place simultaneously is the centralization of capital. It refers to the tendency of existing capital to be concentrated in the hands of fewer and fewer number of larger and larger capitalists. When capital increases the number of capitalists falls. This centralization takes place due to several reasons. There is the tendency for the rate of profit to fall. During the business cycles there is overproduction and huge stocks of goods remain unsold. With the increasing use of machinery and increase in labour productivity, goods become cheaper and cheaper. All these factors give rise to a severe competitive struggle among the producers. Competitive struggle serves as an agent of centralization. It always ends in the ruin of many small capitalists, whose capitals pass into the hands of their conquerors, partly vanish. The strong either crushes or absorbs the weak one. One capitalist kills many. The expropriators are expropriated. The small capitalists who lose in the battle of competition, drop out of the race and join the ranks of the proletariat, with the elimination of the smaller enterprises, competition disappears. The economic scene comes to be dominated either by a single capitalist or a single giant corporation. The credit system further aggravates the process as finally it transforms itself into an immense social mechanism for the centralization of capitals. In the development of capitalist production, concentration and centralization become self-perpetuating.

Capitalism reduces all intermediate classes to the level of the proletariat. According to Marx this would happen as ultimately the proletariat would swallow up all other classes; except the capitalists. Competition would grow more and more severe which would kill many small capitalists. Capitalism would invade every branch of productive activity. This would proletarianize the intermediate classes of artisans, shopkeepers and peasants. Concentration and centralization of capita will lead to a stage of 'Finance capital'. In this stage of capitalist development monopolies will play a predominant role m the economic life, after replacing free competition. There is a merger of bank capital with industrial capital. There is an unholy alliance between monopolist producers and powerful bankers and financial interests. A financial oligarchy is created on the basis of this 'finance capital'. Production is controlled through credit. The monopolist now not only earns profits but super-profits. The capitalist reaches the peak of his parasitic existence. Finally, once finance capital becomes dominant domestically, the financial magnates reach out beyond the boundaries of individual capitalist countries. This stage is known as capitalistic imperialism. This is a stage of international capitalism. At this stage the contradictions created by capitalist production are not limited to' a single country. They assume an international character. Capitalistic imperialism bas been called the highest stage of capitalism by Lenin, because all the contradictions inherent in the system of capitalist production assume their full and ultimate form with it.

According to Marx, the process of accumulation under capitalism inevitably leads to certain contradictions. There is an alienation and exploitation of labour. There is a secular tendency for the rate of profit to decline. Business cycles or economic crises occur regularly and with increasing ferocity. Economic power is concentrated in the hands of fewer and fewer capitalists. The economic life comes to be dominated by giant monopolistic and oligopolistic organizations. There is a steady deterioration in the condition of the workers. Thus inherent contradictions of the capitalist system come to the surface rendering its operation increasingly difficult. Life gradually becomes intolerable. At last there would be so serious a catastrophe that the capitalist system would destroy itself. This will mark the beginning of a revolutionary period during which the working class, at once oppressed and disciplined by its special position in society, will overturn the existing relations of production and establish in their stead higher Socialist relations of production. This show through a socialist revolution the capitalist order would be inevitably transformed into socialism. Thus Marx's economic analysis flows into the sociology of class and socialism. Thus, we find that capitalism will not finally collapse by itself in an automatic mechanical way. The destruction of capitalism must be the work of human agency. The key to social transformation lies in the antagonistic relations between classes. This transformation is brought about by the proletariat with a conscious effort-through violence and revolution.

6

Difference between Classical and Neoclassical Approach

6.1. Difference between Classical and Neoclassical Approach

The difference between classical and orthodox theory Classical and neoclassical theories are distinguished in terms of their themes in analyzing economy, methodology, and value theory. The question of Economics Classical economists investigated two central economic questions: what causes an economy to grow? What determines the distribution of income into its three forms of wages, rent and profit? The classical theme is the accumulation and allocation of surplus output, and therefore their emphasis was on production and on the factors that influence the supply of goods. Neoclassical economics, unlike its classical predecessor, focuses on individual choices, which unavoidably reflect subjective preferences and beliefs, and the allocation of given resources among alternative uses. Indeed, the marginalist revolution is a shift in focus from capital accumulation and growth to utility maximization and production efficiency as the aims of economic policy. This shifted the foundation of economics from production to exchange. On the other hand, the Classical theory paid comparatively little attention to choices of individuals. The classical economists did not believe that there was much of general interest to be said about the preferences of or choices of individuals. Instead, they divided agents into three major classes: capitalists with their capital or stocks of accumulated goods, landlords with their land, and workers with their ability to work.

Neoclassical economics is a term variously used for approaches to economics focusing on the determination of prices, outputs, and income distributions in markets through supply and demand, often mediated through a hypothesized maximization of utility by income-constrained individuals and of profits by cost-constrained firms employing available information and factors of production, in accordance with rational choice theory. Neoclassical economics dominates microeconomics, and together with Keynesian economics forms the neoclassical synthesis which dominates mainstream economics today. Although neoclassical economics has gained widespread acceptance by contemporary economists, there have been many critiques of neoclassical economics, often incorporated into newer versions of neoclassical theory as awareness of economic criteria changes.

6.1.1. Holism vs. Individualism

Methodological holism is a view, according to which properties of individual elements in a complex are taken to be determined by relations they bear to other elements. That is, the whole is more than its parts in the sense that the properties of the whole come from the systematic relational patterns of interaction among its parts and not simply from the aggregation of their atomistic (*i.e.* non-relational) properties. In other words, the social, aggregative implications of individual choices are often unintended. Methodological individualism, however, is the view that all social events must be explained as consequences of choices made by individuals and allows only individuals to be the decision-makers in explaining social phenomena. Accordingly, methodological individualism does not allow explanations which involve non-individualist decision makers such as institutions. Classical economic theories, in general, explained the social phenomena in terms of methodological holism. Their primary interests are social entities such as capitalist society. Moreover, classical economists develop their theory in terms of class such as landlords, workers and capitalists who contest for a larger share of the pie, instead of abstract individuals. In a classical paradigm, the concept of class is more useful than individual agents in analyzing social phenomena. In this way, classical economy has roots in holism, and accommodates social categories like class. Usually the neoclassical economics is firmly grounded on a research program of methodological individualism, which generally posits that individuals with given preferences and endowments, and firms with given technologies, enter the market process as autonomous entities. Indeed, the adherence to the tradition of methodological individualism has an important bearing on the nature of the neoclassical model. For neoclassical economists, society is the collection of individuals. Individual wants, thoughts, and deeds combine to make society what it is. To understand an economy is then to make sense of the aggregate effects of individual wants and acts. Neoclassical theory does this by demonstrating how individuals maximize their material self-interests by utilizing their resources and the available technology in market transactions. What happens in an economy is always explained as the result of individuals acting in this way. The economy, as neoclassical economists theorize it, is the aggregate end product of individuals.

6.1.2. Historical vs. a Historical

In classical economics tradition, historical analysis is called for and mathematical models and statistical investigations are of limited usefulness. The classical economists have made consistent efforts to explain the rise of the capitalist mode of production in terms of historical analysis. They from the beginning have sensed that it was the essence of capitalist production to generate an accumulating surplus, and that the manner of this accumulation was crucially related to the character of the newly emerging social relations of capitalist society. The classical theory of value would be limited to the goods and services that were typical products of competitive capitalism. They regarded the explanation of this historical phenomenon as their primary task. Instead of the characteristic features of a given socioeconomic system, the starting point of neoclassical economics is the a historical individual and therefore the neoclassical theories are expressed in mathematical models which exclude the concept of history. For example, in neoclassical traditions, there are only factors of production, namely, labour, land and capital, each of which receives a reward commensurate with its productivity. There are no essential differences among them. This shows the lack of the notion of class which is unique to the stage of capitalist society. This methodological feature arises from the fact that though the great themes of classical economy are dynamic and developmental, the great themes of neoclassical economy are static and allocation.

Most schools of economic thought have a theory of value at the heart of their paradigm. In addition to identifying the forces that form prices, the theory of value reveals the basic structure of the paradigm. The particular theory of value with which an economist begins is almost invariably a sort of shorthand expression of the basic attitude which the economist is going to adopt towards the phenomena he seeks to analyze and the problem she seeks to solve. In terms of explanations of what determines the value of different commodities, the different approaches have tended to distinguish two notions of value: the cost of producing the commodity reflected by the supply of it, or determined by the utility it gives a consumer, and reflected in the demand for it. The classical economists found value to be determined in production, that is, in the workshop or the factory, not the market place; the value of a bushel of corn, for example, was thought to depend on the costs involved in producing that bushel. Since most of the cost of production could be reduced to labour, this approach was refined into the Labour Theory of Value. They were not, however, very concerned about demand as a determinant of value. Ricardo explicitly excepts goods, like rare paintings, whose price depends solely on their scarcity, from the general principles of value determination through labour time. They did recognize that the usefulness or use-value of a commodity is a preconditions for its having exchange value, but also pointed out that the overall usefulness of commodities had no correlation with their value. The classical theories of value gradually gave way to a perspective in which value was associated with the relationship between the object and the person obtaining the object. The neoclassical "revolution" involved the abandonment of the classical theory of value based on the centrality of labour or work in the production of wealth and its replacement with a preoccupation with the "utility" gained from the consumption of wealth. Neoclassical economists view value in terms of the relationship between costs of production and subjective elements, called supply and demand. They emphasize that prices also depend upon the level of demand, which in turn depends upon the amount of consumer satisfaction provided by individual goods and services. The neoclassical theory of value, therefore, is neoclassical theory of supply and demand behaviour. For neoclassical economics, the value of a commodity is determined only by its subjective scarcity, *i.e.* the degree to which people's desire for that a commodity exceeds it's availability. It is the Marginal Theory of Value or subjective theory of value. Typically, the neoclassical answer to these value questions involves the specification and use of what might be called market analysis. Markets are considered locations or sites where values are determined.

6.1.3. Carl Menger (1840-1921)

Carl Menger is generally considered as the leader of the Austrian school. Menger obtained a doctor's degree from the University of Cracow and joined Austrian civil service. In 1873 he resigned the job to join the University of Vienna as professor. Menger's best-known writings are 'foundations of Political Economy' (Grunsatze). 'Inquiries into the method of social science' and 'On the theory of capital'. Menger's main economic ideas can be considered under the following heads:

(a) Menger's Methodology

When Menger started writing, the academic literature was dominated by the younger Historical school of Germany under the influence of Professor Schmoller. Most German economists believed in inductive and historical method of reasoning and believed that history alone could explain economic phenomena. They held that economic theory had nothing to do with scientific research. They rejected theoretical analysis and deductive method. Some of them even argued that the world 'theory' should be replaced by 'doctrine'. Menger in his writings defended theoretical analysis and deductive methods. He recognized the value of historical research, but he insisted that theoretical analysis and deductive method were also necessary. This difference in opinion between Schmoller and Jvlenger led to a 'Controversy on methods' between the historical and Austrian Schools which was referred to as the "Methodenstreit".

(b) Theory of Goods

Menger held that, in economics, the fundamental thing is the human want. Things which can be placed in the satisfaction of human wants are called utilities. They are usually called goods. An object becomes a good only when it satisfies four conditions, namely:

(i) There must exist a human want.

(ii) The object must possess qualities that can satisfy the want.

(iii) Man must recognize this cause and effect or the want satisfying capacity of the object.

(iv) Man must have command over the thing so that it may be used to satisfy want.

Having explained the cause and effect relation between goods and wants, Menger went further to classify different types of goods. He divided goods into different classes or orders based on their nearness to the consumer. Consumer's goods or 'goods of the first order' are defined as those which have the capacity of satisfying immediate needs or wants. Producers' goods come in the later order (like second order goods, third order goods etc). Producers' goods are capable of producing consumer' goods. For example, bread is a consumers' good or first order good since it satisfies an immediate human want, hunger. But flour with which the bread is made is a second order good since it satisfies a want which comes only next to the immediate want hunger. Also note that from the second order good or producers' good flour we are able to make bread which is a consumers' good or first order good and the producers' good (flour) derives its value from the consumers' good (bread). Similarly, wheat is a third order good. Menger also used this idea to explain the concept of complimentary goods. Complimentary goods are those which appear in the later order. For example, the chain of complementary relation runs as follows. The demand for bread creates a demand for flour, which in turn creates a demand for wheat etc. Manger also distinguished between free goods (non-economic goods) and economic goods. Economic goods are those goods the requirement of which exceeded their supply. Non-economic goods or free goods are those which are unlimited in supply (*e.g.* air, water).

(c) Menger's Value Theory

Menger further analyzed how the values of the goods are determined. He held that value is a subjective concept. According to him the origin of all economic value is the relation between a human want and the available economic goods capable of satisfying the want. Hence an increase or decrease in the quantity supplied altered the value of the commodity. His main argument was that as quantity of a good increased, the additional unit had less want-satisfying power. Menger further points out that value depends on the significance attached to a thing by different persons based on their psychological judgments'. For example, if person 'A' thinks that good 'al' is of superior quality than good 'a2', he will attach greater value to 'al'. On the other hand, if another persons'B' thinks that good 'a2' is of superior quality to good 'al', then he will attach greater value to it. However, the above principle cannot be used to measure value of goods of a higher order of the producer's goods. They have to be imputed based on the consumers' good which they help to produce. Thus, for example, the value of flour depends on bread. This is called the theory of imputation and it is considered as a major contribution by Menger, since it is the first attempt to explain distribution scientifically.

(c) Wealth

Menger defined wealth as the sum total of "economic goods at the command of an economizing individual". He called this 'individual wealth'. Just as an individual possessed wealth to satisfy his or her requirements, in the same way, states, provinces, communities and association also had large quantities of economic goods, which he called public wealth. The national wealth is the totality of the individual wealth and public wealth.

(d) Theory of Exchange

According to Menger, exchange enabled individuals to satisfy their wants better. In an exchange both buyer and the seller gained in others of satisfaction. Menger felt that, just like individuals, nations can also gain by exchange. He pointed out that two nations - one engaged mainly in agriculture and the other in industry - would satisfy their needs in a better manner by means of exchange.

(e) Theory of Money

Menger believed that money evolved due to the inconveniences of barter. He differentiated between the transaction and precautionary motives behind holding money. He held that metals are more suitable for the preservation of value than other commodities. He suggested that the value of money might be stabilized by regulating the production of gold and the amount of notes in circulation.

(f) Use - Value and Exchange - Value

Menger said that a good is said to have use-value if it sincerely satisfies the needs of the individual. On the other hand, a good is said to have exchange-value, if it indirectly satisfies the needs of the individual. The same good may have use-value and exchange value at different times. For example toys had use value for a person in his childhood, but they have only exchange value in his adult hood. To conclude, let us see some of the major limitations of Menger's theories. He overemphasized demand to the point of insisting that cost of production had nothing to do with value. This is true only in the very short run. In the long run, we cannot assume cost to be constant and supply will be greatly influenced by cost. Another criticism is that his theory of imputation is difficult to put on practice. Yet nanother criticism is that he could not develop a complete theory of capital and price.

6.1.4. Leon Walras (1835-1910)

Leon Walras, a French economist, was the son of Augusta Walras, who was an economist. Walras was a professor of political economy at the University of Lausanne, Switzerland. Walras is considered as the founder of the Lausanne school or the mathematical school of economics. He was a born thinker and is considered as the greatest of pure theorists. His important contributions are 'Elements of pure Economies', Theory of exchange', Theory of production', and 'Studies in applied Economies'. Following are his important contributions.

As a social reformer, Walras recommended threefold concepts; the nationalization of land by the State, the abolition of taxation and the securing of competitive conditions in industry so that the "idealized" capitalism of his Elements would be attained. He believed that in this, he had found the "scientific solution" to humanity's problem. Thus convinced he had four of his followers write to the Nobel Committee in 1905 nominating him for a Nobel Prize. Walras's sole academic job was as an economics professor at the University of Lausanne in Switzerland. This location was not ideal: because the dominant thinking in economics at the time was in Britain, it was difficult for Walras to affect the rest of the profession. Also, because his students were more interested in becoming lawyers than in becoming economists, Walras did not have disciples. Although his impact on economics was limited during his lifetime, it has been much greater since the 1930s. Historian of economic thought Mark Blaug wrote that Walras"may now be the most widely-read nineteenth century economist after Ricardo and Marx."

- Walras' methodology was subjectivism and marginalism. He made use of mathematical ideas in his analysis. Along with Gossen, Jevons and Menger, Walras should be given credit for developing marginal utility analysis. However Walras developed his analysis independent of others. For example, instead of the term "marginal utility", while Gossen used "value of the last atom" and Jevons used "final degree of utility". Gossen used the term "rarete" which he defined as intensity of last want satisfied. He held that value of things is proportional to the satisfaction they bring us. He further held that exchange value of a commodity is in proportion to 'rarete'.
- On the basis of the marginal utility theory, Walras developed the concept of General equilibrium. His analysis s based on the following assumptions (i) There is free entry and competition (ii) consumers' tastes, technology and factors of production remain constant. Walras divided 'the entire economy into two markets namely, a product market and a factor market. In the product market the sellers are business firms who produce the goods and the buyers are the households who demand the goods. In the factor market, the business firms become the buyers while the households who own factors of production become sellers. According to Walras, the buyers and the sellers go to their respective markets and each is guided by the desire to maximize their utility. When the prices at which the sellers are willing to sell equal the prices at which the buyers are willing to buy, general equilibrium is achieved. In other words when buyers and sellers maximize their utility, general equilibrium is attained. '
- In his theory of capital, Walras made a clear distinction between capital goods and consumption goods. Capital goods are desired by people for the income they earn while consumption goods are demanded for want satisfaction. He held that interest is the price for the services of capital.
- At several instances in the writings, Walras has mentioned the functions of the state, which are (i) To maintain the system of free competition through legislation (ii) Intervention of the state in matters like education, speculative activities, monetary policy and labour market (iii) The maintenance of justice and security.
- Walras viewed his general economic equilibrium in the form of a system of simultaneous equations. He attempted to reveal the linking many markets of which the economy is made up. In his system of equations, he started with two different markets: products markets and markets for productive services. He concluded that the consumers are the buyers in the products markets, while they are sellers in the markets for productive services. At the same time, the business firms are buyers in the markets for productive services and sellers in the products markets.In the demand equations, the quantity demanded of a product is related not only to the price of the product but also to the priced of all other products and of the productive services. The prices of productive services actually are the consumers' income, so affecting their demand for the products. Also, the prices of

other products either substitutes or complementary goods influence consumers' demand for a particular product. His demand equations also are derived from utility functions, which are maximized when for each consumer prices are proportional to marginal utilities. Additionally, maximizing utility functions constrained by budget equations. According to Walras's law, whenever one of the markets is clear, then all markets are clear. Walras also introduced the concept of a backward sloping supply curve, which could intersect the demand curve more than once, thus yield multiple equilibria. He suggested the possibility that consumers prefer leisure to work. Later on, Walras also expanded his system of equations to account for saving, the formation of capital, and the holding of money. He extended his analysis of the demand of consumer goods to saving and to money. In terms of saving, Walras considered only from the utility of perpetuity, no difference in time preference. His demand of money equation also expressed under certainty.

6.2. Friedrich List

List was born in Reutlingen, Württemberg. Unwilling to follow the occupation of his father, who was a prosperous tanner, he became a clerk in the public service, and by 1816 had risen to the post of ministerial under-secretary. In 1817, he was appointed professor of administration and politics at the University of Tübingen, but the fall of the ministry in 1819 compelled him to resign. As a deputy to the Württemberg chamber, he was active in advocating administrative reforms. Arriving in the United States in 1825, he settled in Pennsylvania, where he became an extensive landholder. He first engaged in farming, but soon switched to journalism and edited a German paper in Reading. He was active in the establishment of railroads. It was in America that he gathered from a study of Alexander Hamilton's work the inspiration which made him an economist of his pronounced 'National System 'views which found realization in Henry Clay's American System. In 1827 he published a pamphlet entitled *Outlines of a New System of Political Economy*, in which he defended the doctrine of protection. The discovery of coal on some land which he had acquired made him financially independent. In 1830, he was appointed United States consul at Hamburg, but on his arrival in Europe in he found that the Senate had failed to confirm his appointment. After residing for some time in Paris, he returned to Pennsylvania. He next settled in Leipzig in 1833, where for some time he was U. S. consul. He was a journalist in Paris from 1837 to 1843. He wrote several letters for the *Augsburg AllgemeineZeitung*, which were published in 1841 in a volume under the title of *Das nationale System der politischenOekonomie*.In 1843 he established the Zollvereinsblatt in Augsburg, a newspaper in which he advocated the enlargement of the customs union (German: *Zollverein*), and the organization of a national commercial system.He strongly advocated the extension of the railway system in Germany. The development of the Zollverein to where it unified Germany economically was due largely to his enthusiasm and ardour.In 1841, his ill health had led him to decline an offer to edit the *RheinischeZeitung*, a new Cologne paper of liberal views, and Karl Marx took the post. He visited Austria and Hungary in 1844. In 1846, he visited England with a view to forming a commercial alliance between that country and Germany, but was unsuccessful.His latter days were darkened by many misfortunes; he lost much of his American property in a financial crisis, ill-health also overtook him, and killed himself on 30 November 1846. He opposed the cosmopolitan principle in the contemporary economical system and the absolute doctrine of free trade which was in harmony with that principle, and instead developed the infant industry argument, to which he had been exposed by Hamilton and Raymond. He gave prominence to the national idea and insisted on the special requirements of each nation according to its circumstances and especially to the degree of its development. He famously doubted the sincerity of calls to free trade from developed nations, in particular Britain: Any nation which by means of protective duties and restrictions on navigation has raised her manufacturing power and her navigation to such a degree of development that no other nation can sustain free competition with her, can do nothing wiser than to throw away these ladders of her greatness, to preach to other nations the benefits of free trade, and to declare in penitent tones that she has hitherto wandered in the paths of error, and has now for the first time succeeded in discovering the truth.

6.2.1. Thornstein Veblen

Thornstein Veblen was an American economist and sociologist, and leader of the institutional economics movement. Besides his technical work he was a popular and witty critic of capitalism, as shown by his best known book The Theory of the Leisure Class (1899). Veblen is famous in the history of economic thought for combining a Darwinian evolutionary perspective with his new institutionalist approach to economic analysis. He combined sociology with economics in his masterpiece The Theory of the Leisure Class (1899) where he argued that there was a fundamental split in society between those who make their way via exploitation and those who make their way via industry. In early barbarian society this is the difference between the hunter and the gatherer in the tribe, but as society matures it is the difference between the landed gentry and the indentured servant. The titular manifestation of those with the power of exploit is the "leisure class" which is defined by its lack of productive economic activity and its commitment to demonstrations of idleness. As Veblen describes it, as societies mature, conspicuous leisure gives way to "conspicuous consumption", but both are performed for the sole purpose of making an invidious distinction based on pecuniary strength, the demonstration of wealth being the basis for social status.

Veblen was sympathetic to state ownership of industry, but he had a low opinion of workers and the labour movement and there is disagreement about the extent to which his views are compatible with Marxism, socialism or anarchism. As a leading intellectual of the Progressive Era, he made sweeping attacks on production for profit, and his stress on the wasteful role of consumption for status greatly influenced socialist thinkers and engineers who sought a non-Marxist critique of capitalism. Fine (1994) writes that economists at the time complained that his ideas, while brilliantly presented, were crude, gross, fuzzy, and imprecise; others complained that he was a wacky eccentric. Scholars continue to debate what exactly he meant in his convoluted, ironic and satiric essays; he made heavy use of examples of primitive societies, but many examples were pure invention. He studied economics under John Bates Clark, a leading neoclassical economist, but rejected his ideas. He did his graduate work at Johns Hopkins University under Charles Sanders Peirce, the founder of the pragmatist school in philosophy, and at Yale University under laissez-faire proponent William Graham Sumner. Veblen is best known for his book *The Theory of the Leisure Class,* which introduced the term "conspicuous consumption" (referring to consumption undertaken to make a statement to others about one's class or accomplishments). This term, more than any other, is what Veblen is known for. Some types of luxury goods, such as high-end wines, designer handbags, and luxury cars, are considered Veblen goods, in that decreasing their prices decreases people's preference for buying them because they are no longer perceived as exclusive or high-status products. Similarly, a price increase may increase that high status and perception of exclusivity, thereby making the good even more preferable. Often such goods are no better or are even worse than their lower priced counterparts.

At the other end of the spectrum, where luxury items priced equal to non-luxury items of lower quality, all else being equal more people would buy the luxury items, even though a few Veblen-seekers would not. Thus, even a Veblen good is subject to the dictum that demand moves conversely to price, although the response of demand to price is not consistent at all points on the demand curve. Veblen developed a 20th-century evolutionary economics based upon Darwinian principles and new ideas emerging from anthropology, sociology, and psychology. Unlike the neoclassical economics that was emerging at the same time, Veblen described economic behavior as socially determined and saw economic organization as a process of ongoing evolution. Veblen strongly rejected any theory based on individual action or any theory highlighting any factor of an inner personal motivation. Such theories were according to him "unscientific." This evolution was driven by the human instincts of emulation predation, workmanship, parental bent, and idle curiosity. Veblen wanted economists to grasp the effects of social and cultural change on economic changes. In The Theory of the Leisure Class, the instincts of emulation and predation play a major role. People, rich and poor alike, attempt to impress others and seek to gain advantage through what Veblen coined "conspicuous consumption" and the ability to engage in "conspicuous leisure." In this work Veblen argued that consumption is used as a way to gain and signal status. Through "conspicuous consumption" often came "conspicuous waste," which Veblen detested.

In The Theory of Business Enterprise, which was published in 1904 during the height of American concern with the growth of business combinations and trusts, Veblen employed his evolutionary analysis to explain these new forms. He saw them as a consequence of the growth of industrial processes in a context of small business firms that had evolved earlier to organize craft production. The new industrial processes impelled integration and provided lucrative opportunities for those who managed it. What resulted was, as Veblen saw it, a conflict between businessmen and engineers, with businessmen representing the older order and engineers as the innovators of new ways of doing things. In combination with the tendencies described in The Theory of the Leisure Class, this conflict resulted in waste and "predation" that served to enhance the social status of those who could benefit from predatory claims to goods and services. Veblen generalized the conflict between businessmen and engineers by saying that human society would always involve conflict between existing norms with vested interests and new norms developed out of an innate human tendency to manipulate and learn about the physical world in which we exist. He also generalized his model to include his theory of instincts, processes of evolution as absorbed from

Sumner, as enhanced by his own reading of evolutionary science, and Pragmatic philosophy first learned from Peirce. The instinct of idle curiosity led humans to manipulate nature in new ways and this led to changes in what he called the material means of life. Because, as per the Pragmatists, our ideas about the world are a human construct rather than mirrors of reality, changing ways of manipulating nature lead to changing constructs and to changing notions of truth and authority as well as patterns of behavior (institutions). Societies and economies evolve as a consequence, but do so via a process of conflict between vested interests and older forms and the new. Veblen never wrote with any confidence that the new ways were better ways, but he was sure in the last three decades of his life that the American economy could, in the absence of vested interests, have produced more for more people. In the years just after World War I he looked to engineers to make the American economy more efficient.

In addition to The Theory of the Leisure Class and The Theory of Business Enterprise, Veblen's monograph "Imperial Germany and the Industrial Revolution", and his many essays, including "Why is Economics Not an Evolutionary Science," and "The Place of Science in Modern Civilization," remain influential. Veblen did not reject economists' answers to the questions they posed; he simply thought their questions were too narrow. Veblen wanted economists to try to understand the social and cultural causes and effects of economic changes. What social and cultural causes were responsible for the shift from hunting and fishing to farming, for example, and what were the social and cultural effects of this shift? Veblen was singularly unsuccessful at getting economists to focus on such questions. His failure may explain the sarcastic tone his writing took toward his fellow economists. Veblen had to struggle to stay in academia. In the late nineteenth century many universities were affiliated in a substantial way with churches. Veblen's skepticism about religion and his rough manners and unkempt appearance made him unattractive to such institutions. As a result, from 1884 to 1891 Veblen lived on the largesse of his family and his wife's family. His big break came in 1892 when the newly formed University of Chicago hired his mentor, J. Laurence Laughlin, who brought Veblen with him as a Teaching Assistant.

Veblen later became the managing editor of the *Journal of Political Economy*, which was and is edited at the University of Chicago. Veblen spent fourteen years at Chicago and the next three at Stanford. He died in obscurity in 1929.

6.3. Important Economic Ideas of Veblen

☆ Veblen held strong critical views on the postulates of economics as it had developed over the years. Veblen was not satisfied with the methods and doctrines of the classical economists and marginalists. He held that economics is fundamentally unsound and that economists were not modern. This was mainly because of two features of the economic science; one is its hedonistic psychology which is based upon the pleasure-pain calculus and two, the assumption of a 'meliorative trend' or a belief that the world may be made better by human effort. He also pointed out the limitations of the marginal utility principle. After criticizing the existing methodology of analysis, Veblen developed his own methodology which he called the 'evolutionary approach.' According this approach, the human behaviour ultimately develop into the form of an institution. In other words, the human behaviour gradually become an accepted pattern for the entire society and develop into habits and customs which are called institutions. Thus, he held that, the present institutions evolved as the result of a gradual process of cultural development being undertaken during the last many centuries.

- Significance of Institutionalism: Veblen's thought is centered around the institution, its evolution and its reaction upon man. Veblen held that a critical examination of the origin and causes of an institution is necessary to understand economic phenomena. Veblen stated that the growth of institutions was mainly due to two reasons namely material environment and propensities of human nature. Veblen held that capitalism has two important institutions, namely industry and business. Around these two institutions, several other small institutions like competition, private property, inheritance, banks, money, price system, credit, profit etc have developed. All these institutions are developed on the basis of habits and thoughts of human behaviour through the evolutionary process. He believed that the process of institutional evolution would continue, and the social institutions would change with changing habits of the society. However he admitted that factors like population, idle curiosity, physical environment, race mixing etc have an influence on the development of institutions.
- The Theory of Leisure Class: Veblen's institutionalism and evolutionary economics rest on 'class conflict' which is expressed in his theory of leisure class. Veblen begins with the assumption that initially property and technological methods were the most important institutions of capitalism. In the beginning, technological information was open to all classes. But as industries developed, production exceeded over requirements and a surplus appeared. Certain classes or groups of the society managed to take away this surplus and started to live in leisure. They are referred to as the 'leisure class.' The leisure class was interested only in money making and living in comfort. On the other hand, there is the productive class which consist of those who work and produce. Ultimately, this will lead to conflicts between the two classes in the society, which may lead to business crises.
- In his work theory of Business Enterprise Veblen gave his views regarding depression and unemployment. Veblen in this book first gave a theoretical

interpretation of the working of large scale industries and other small scale institutions of American capitalist system. He felt that though the use of machines had increased the efficiency of the industries, the advances of these changes was not evenly distributed in the society. This was because most of the industrialists were motivated by larger profits which led to cut-throat competition and monopolies. This resulted in waste of resources, depression and unemployment. He held that the large-scale industries in the existing capitalism were parasites on the community.

☆ Veblen has also given an analysis of trade cycles in which he held that business fluctuations are the result of business enterprise. During times of prosperity, the businessmen, in their effort to maximize profit tend to indulge in excessive use of loan and adopt such profit making measures which reduce efficiency. However, the increased borrowing from bank results in high interest rates and the increased production will result in high cost of inputs and labour. Thus in the later stages of prosperity, profits decrease and business becomes costlier. This ultimately results in overproduction and hence depression. Veblen further believed that the businessmen not only produce economic crisis through their excessive use of credit, but also attempt to intensify the crisis.

6.4. Vilfredo Pareto (1848–1923)

Pareto was an Italian engineer, sociologist, economist, and philosopher. Born in Paris to Italian exiles, Pareto moved to Italy to complete his education in mathematics and literature. After graduating from the Polytechnic Institute in Turin in 1869, he applied his prodigious mathematical abilities as an engineer for the railroads. Throughout his life Pareto was an active critic of the Italian government's economic policies. He published pamphlets and articles denouncing protectionism and militarism, which he viewed as the two greatest enemies of liberty. Although he was keenly informed on economic policy and frequently debated it, Pareto did not study economics seriously until he was forty-two. In 1893 he succeeded his mentor, Leon Walras, as chair of economics at the University of Lausanne. His principal publications are *Coursd'économiepolitique*(1896–1897), Pareto's first book, which he wrote at age fortynine; and Manual of Political Economy (1906).

He made several important contributions to economics, particularly in the study of income distribution and in the analysis of individuals' choices. His legacy as an economist was profound. Partly because of him, the field evolved from a branch of social philosophy as practiced by Adam Smith into a data intensive field of scientific research and mathematical equations. His books look more like modern economics than most other texts of that day: tables of statistics from across the world and ages, rows of integral signs and equations, intricate charts and graphs. He introduced the concept of Pareto efficiency and helped develop the field of microeconomics. He also was the first to discover that income follows a Pareto distribution, which is a power law probability distribution. The pareto principle was named after him and built on observations of his such as that 80% of the land in Italy was owned by 20% of the population. He also contributed to the fields of sociology and mathematics. Pareto is best known for two concepts that are named after him. The first and most familiar is the concept of Pareto optimality. A Pareto-optimal allocation of resources is achieved when it is not possible to make anyone better off without making someone else worse off. The second is Pareto's law of income distribution. This law, which Pareto derived from British data on income, showed a linear relationship between each income level and the number of people who received more than that income.Pareto found similar results for Prussia, Saxony, Paris, and some Italian cities. Although Pareto thought his law should be "provisionally accepted as universal," he realized that exceptions were possible; as it turns out, many have been found.

Pareto is also known for showing that the assumption that the utility of goods can actually be measured is not necessary to derive any of the standard results in consumer theory. Simply by being able to rank bundles of goods, consumers would act as economists had said they would In his later years Pareto shifted from economics to sociology in response to his own change in beliefs about how humans act. He came to believe that men act non logically, "but they make believe they are acting logically." The most widely-used concept in theoretical welfare economics is "Pareto optimality" (also known as "Pareto efficiency"). An allocation is Pareto-optimal iff it is impossible to make at least one person better off without making anyone else worse off; a Pareto improvement is a change in an allocation which makes someone better off without making anyone else worse off. Given an initial allocation of goods among a set of individuals, a change to a different allocation that makes at least one individual better off without making any other individual worse off is called a Pareto improvement. An allocation is defined as "Pareto efficient" or "Pareto optimal" when no further Pareto improvements can be made. Pareto efficient allocation is one for which each agent is as well off as possible, given the utilities of the other agents.""Better" and "worse" are based purely upon subjective preferences which can be summarized in a "utility function," or ordinal numerical index of preference satisfaction. It is commonly accepted that outcomes that are not Pareto efficient are to be avoided, and therefore Pareto efficiency is an important criterion for evaluating economic systems and public policies. If economic allocation in any system is not Pareto efficient, there is potential for a Pareto Improvement–an increase in Pareto efficiency: through reallocation, improvements can be made to at least one participant's well-being without reducing any other participant's well-being. It is important to note, however, that a change from an inefficient allocation to an efficient one is not necessarily a Pareto improvement. Thus, in practice, ensuring that nobody is disadvantaged by a change aimed at achieving Pareto efficiency may require compensation of one or more parties. For instance, if a change in economic policy eliminates a monopoly and that market subsequently becomes competitive and more efficient, the monopolist will be made worse off. However, the loss to the monopolist will be more than offset by the gain in efficiency. This means the monopolist can be compensated for its loss while still leaving a net gain for others in the economy, a Pareto improvement. In real-world practice, such compensations have unintended consequences. They can lead to incentive distortions over time as agents anticipate such compensations and change their actions accordingly. Under certain idealized conditions, it can be shown that a system of free markets will lead to a Pareto efficient outcome. This is called the first welfare theorem. It was first demonstrated mathematically by economists Kenneth Arrow and Gérard Debreu. However, the result only holds under the restrictive assumptions necessary for the proof (markets exist for all possible goods so there are no externalities, all markets are in full equilibrium, markets are perfectly competitive, transaction costs are negligible, and market participants have perfect information). In the absence of perfect information or complete markets, outcomes will generically be Pareto inefficient. While initially it might seem that every situation is necessarily Pareto optimal, this is not the case. True, if the only good is food, and each agent wants as much food as possible, then every distribution is Pareto optimal. But if half of the agents own food and the other half own clothes, the distribution will not necessarily be Pareto optimal, since each agent might prefer both more food and fewer clothes or vice versa. In the final analysis, welfare economists' attempt to provide a value-free or at least value-minimal justification of the state fails quite badly. Nevertheless, economic analysis may still inform more substantive moral theories: pareto optimality, for example, is a necessary but not sufficient condition for a utilitarian justification of the state.

6.4.1. Pareto's Important Ideas may be Summarized as Follows

- In his methodology, Pareto treated economics as a pure economic science which will be free from ethical considerations. He used the quantitative method to develop economics as a mathematical discipline because he held the view that mathematics was the only subject which could provide exactness to economics.
- Pareto contributed substantially to the development of new Welfare Economics. He developed an optimality criterion by which social welfare resulting from an economic action can be measured. In this respect he held that utility was a psychological phenomenon, and was, therefore, not measurable. So he replaced the cardinal approach by the ordinal approach, on which the indifference curve analysis was based. Based on indifference curve analysis he developed his optimality condition which is as follows. "Any change that makes at least one individual better off and no one worse off is an improvement in social welfare. Conversely, changes that make no one better off and at least one worse off is a decrease in social welfare".
- Another important contribution by Pareto was the law of income distribution. After analyzing data on income from different nations, he arrived at the conclusion that the pattern of income distribution was basically the same in different countries and for different times. According to him inequality was a result of unequal human capacities. He suggested an increase in projection and income at a faster rate than that of population.

7
Arthur Cecil Pigou (1877–1939)

7.1. Arthur Cecil Pigou (1877–1939)

Arthur Cecil Pigou was born in the family home of his mother on November 18, 1877, at Ryde, in the Isle of Wight. He was the eldest son of Clarence and Nora Pigou. His father came from the Huguenot line and his mother's family came from a line that had won fame and fortune in Irish administration. The pride and background of Pigou's family helped to push him along his path later in life. His abilities in academics gained him an entrance scholarship to the school. Athletics was also one of Pigou's strong points. His talents in sports allowed him to be approved of by many at a time in history where athletics was looked at as being more important than academics. He ended his stay at Harrow as head of the school.Pigou's work is notable in two areas: welfare economics and the theory of unemployment. As in his major work The Economics of Welfare Pigou was strongly influenced by his former teacher Alfred Marshall. Afterwards, he went to King's College, Cambridge as a history scholar. There, he came to economics though the study of philosophy and ethics under the Moral Science Tripos. He studied economics under Alfred Marshall, and in 1908 Pigou was elected professor of Political Economy at Cambridge as Marshall's successor. He held the post until 1943.One of his early acts was to provide private financial support for John Maynard Keynes to work on probability theory. Pigou and Keynes had great affection and mutual regard for each other and their intellectual differences never put their personal friendship seriously in jeopardy.Pigou was a devoted expositor of Marshallian economics while he held the Cambridge chair. His most important work was published in 1912 as Wealth and Welfare, but was expanded to become the better known The Economics of Welfare in 1920. He became a Fellow of the British Academy in 1927.Pigou pioneered welfare economics with his concerns for justice and the protection of the interests of the poor. These views were rejected by John Maynard Keynes. Pigou retaliated by producing a severe review of Keynes' book. Despite their academic differences they remained firm friends. Later, Pigou began to appreciate the ideas of Keynes, acknowledging that he had come with the passage of time to feel that he had failed earlier to appreciate some of the important things that Keynes was trying to say. Pigou gave up his professor's chair in 1943, but remained a Fellow of Kings College until his death. In his later years, he gradually became more of a recluse, emerging occasionally from his rooms to give lectures or take a walk.Pigou died in 1959 in Cambridge.

Pigou was of the view that the main aim of economic science is to increase social welfare. In that case, economics would be realistic and more beneficial to the society. In his definition, Pigou defines economics as a study of economic welfare which is that part of social welfare that can be brought directly or indirectly into relation with the measuring rod of money". He, therefore, considers economics as a normative science or welfare economics. Pigou's major work, Wealth and Welfare (1912) and Economics of Welfare (1920), developed Alfred Marshall's concept of externalities, costs imposed or benefits conferred on others that are not taken into account by the person taking the action.Pigou attributed welfare gains to the greater marginal utility a dollar of income had for the poor compared to the rich; a transfer of income from rich to poor increased total utility that could also be defined as increased "quality of life."Pigou also argued that welfare gains came from improving the quality of the work force through changes in the distribution of income or by improved working conditions. He argued that the existence of externalities was sufficient justification for government intervention. The reason was that if someone was creating a negative externality, such as pollution, he would engage in too much of the activity that generated the externality. Someone creating a positive externality, say, by educating himself and thus making himself more interesting to other people, would not invest enough in his education because he would not perceive the value to himself as being as great as the value to society. To discourage the activity that caused the negative externality, Pigou advocated a tax on the activity. To encourage the activity that created the positive externality, he advocated a subsidy. These are now called Pigovian (or Pigovian) taxes and subsidies. Pigou's approach came under attack from Lionel Robbins and Frank Knight. The New Welfare Economics that arose in the late 1930s dispensed with much of Pigou's analytical toolbox. Later, the Public Choice theorists rejected Pigou's approach for its naive "benevolent despot" assumption. Finally, Nobel Laureate Ronald Coase demonstrated that efficient outcomes could be generated without government intervention when property rights are clearly defined. Coase presents his case in the article "The Problem of Social Cost" (1960).To explain this alternative let us continue with the paper mill example. There is a second approach likely taken. In this line of thinking the economist considers the paper mill and others who wish to consume or enjoy water quality as part of a competitive market where people bargain for the use of rights to scarce property. This analysis has nothing to do with polluters' imposing cost on society, but everything to do with competing demands for use of an asset. If rights to the asset are defined and assigned to members of the river-basin community, then those planning to build the paper mill must bargain with the right holders to determine just how much, if any, waste will discharge into the river. If the rights are held by the mill, then the existing communities along the river must bargain with the mill owner for rights to water quality. Again, bargaining determines the amount of discharge to the river.

This approach relies on the work of Ronald Coase (1960). Using this framework, an economist might recommend a meeting of the mill owners and others who have access to the river. After organizing the parties, negotiations would ensue. If existing river users owned water-quality rights, the mill would have to buy the rights in order to discharge specified amounts of waste. If the mill had the right to pollute, existing river users would have to buy water quality from the mill, paying the mill to limit its discharges. In other words, Pigouvian taxes do embody the important principle that polluters should pay for the damages they inflict on society. But in both law and economics, a more conservative analysis has gained popularity. Legal scholar Ronald Coase argued that taxes and regulation might be unnecessary, since under some circumstances polluters and those harmed by pollution could engage in private negotiation to determine the appropriate compensation. While Pigou's examples of externalities often involved simultaneous harms to large numbers of people, Coase's examples tended to be localized, individual nuisances, where one person's behavior disturbed the immediate neighbors. The image of environmental externalities as localized nuisances serves to trivialize the real problems of widespread, collective threats to health and nature. Creative alternative readings of Coase have been suggested at times, but the dominant interpretation of his work has provided an intellectual basis for the retreat from regulation.

7.1.1. Theory of Unemployment

Pigou's classical theory of unemployment (Pigou 1933) is based on two fundamental postulates, namely:The wage is equal to the marginal product of labour.That is to say, the wage of an employed person is equal to the value which would be lost if employment were to be reduced by one unit (after deducting any other costs which this reduction of output would avoid); subject, however, to the qualification that the equality may be disturbed, in accordance with certain principles, if competition and markets are imperfect. The utility of the wage when a given volume of labour is employed is equal to the marginal disutility of that amount of employment. That is to say, the real wage of an employed person is that which is just sufficient (in the estimation of the employed persons themselves) to induce the volume of the actually forthcoming labour; subject to the qualification that the equality for each individual unit of labour may be disturbed by combination between employable units analogous to the imperfections of competition which qualify the first postulate. Disutility here must be understood to cover every kind of reason which might lead a man, or a body of men, to withhold their labour rather than accept a wage which had to them a utility below a certain minimum. This second postulate is compatible with what may be called "frictional" unemployment. For a elastic interpretation of it, we must legitimately allow for various inexactnesses of adjustment which stand in the way of continuous full employment. For example, unemployment due to a temporary loss of balance between the relative quantities of specialized resources as a result of miscalculation or intermittent demand; or to time-lags consequent on unforeseen changes; or to the fact that the changeover from one employment to another cannot be effected without a certain delay, so that there will always exist in a non-static society a proportion of resources unemployed "between jobs."In addition to "frictional" unemployment, the postulate is also compatible with "voluntary" unemployment due to the refusal or inability of a unit of labour, as a result of legislation or social practices or of combination for collective bargaining or of slow response to change or of mere human obstinacy, to accept a reward corresponding to the value of the product attributable to its marginal productivity. But in his thinking, these two categories of "frictional" unemployment and "voluntary" unemployment are considered comprehensive. The classical postulates do not admit of the possibility of the third category, which we might define as "involuntary" unemployment. Subject to these qualifications, the volume of employed resources is duly determined, according to the classical theory, by the two postulates. The first gives us the demand schedule for employment, the second gives us the supply schedule; and the amount of employment is fixed at the point where the utility of the marginal product balances the disutility of the marginal employment. From this it follows that there are only four possible means of increasing employment: An improvement in organization or in foresight which diminishes "frictional" unemployment. A decrease in the marginal disutility of labour, as expressed by the real wage for which additional labour is available, so as to diminish "voluntary" unemployment. An increase in the marginal physical productivity of labour in the wage-goods industries (to use Pigou's convenient term for goods upon the price of which the utility of the money-wage depends); or an increase in the price of non-wage-goods compared with the price of wage-goods, associated with a shift in the expenditure of non-wage-earners from wage-goods to non-wage-goods.

7.1.2. The Pigou Effect

What is now known as the Pigou effect was first popularized by Pigou in 1943. The term refers to the stimulation of output and employment caused by increasing consumption due to a rise in real balances of wealth, particularly during deflation. Pigou had proposed the link from balances to consumption earlier, Gottfried Haberler having made a similar objection the year after the publication of John Maynard Keynes' General Theory. In fact, Haberler in 1937 and Pigou in 1943 both showed that a downward wage-price spiral had the effect of increasing real money balances. As price declines drove up the value of the existing money supply, the increase in real money balances would at some point satisfy savings desires and result in a resumption of consumption. Wealth was defined by Pigou as the sum of the money supply and government bonds divided by the price level. He argued that Keynes' General theory was deficient in not specifying a link from "real balances" to current consumption, and that the inclusion of such a "wealth effect" would make the economy more "self correcting" to drops in aggregate demand than Keynes predicted. Because the effect derives from changes to the "Real Balance," this critique of Keynesianism is also called the Real Balance effect. Pigou later dismissed his "Pigou effect" or "real balance effect" as an academic exercise, because a government would not employ a downward wage-price spiral as a means of increasing the real money supply. In contrast, Karl Polanyi recognized the real world policy implications of the real balance effect. He dismissed the wage price flexibility discussion as irrelevant and stated the "Pigou effect" in terms of constant prices and increases in the nominal stock of money. In Polanyi's approach, the policy issue is not obscured by adverse effects on expectations caused by price level declines. All this, moreover, has its reverse side. In an exchange economy everybody's money income is somebody else's cost. Every increase in hourly wages, unless or until compensated by an equal increase in hourly productivity, is an increase in costs of production. An increase in costs of production, where the government controls prices and forbids any price increase, takes the profit from marginal producers, forces them out of business, and means a shrinkage in production and a growth in unemployment. Even where a price increase is possible, the higher price discourages buyers, shrinks the market, and also leads to unemployment. If a 30 percent increase in hourly wages all around the circle forces a 30 percent increase in prices, labour can buy no more of the product than it could at the beginning; and the merry go-round must start all over again.

7.2. Elasticity of Demand for Labour

An important factor in this analysis is the elasticity in the demand for labour. In this case, elasticity is defined:, e = (percentage change in employment) / (percentage change in wage). Elasticity is the percentage change in quantity (in this case employment) divided by the percentage change in price (or wage.) The labour elasticity should actually be defined in negative numbers. For the sake of simplification we shall use the positive coefficients here as well. For example, an elasticity coefficient of two shows that the labours force responds a great deal to a change in wage. If, on the other hand, a ten percent change in wage causes only a five percent change in employment, the elasticity coefficient will be only one-half. Economists would say in this case that demand is inelastic. Demand is inelastic whenever the elasticity coefficient is less than one. When it is greater than one, economists say that demand is elastic. While analyzing the elasticity of demand for labour, Paul H. Douglas in America from analyzing a great mass of statistics and Pigou in England, by almost purely deductive methods, arrived independently at the conclusion that the elasticity of the demand for labour is somewhere between three and four. This means, in less technical language, that "a one percent reduction in the real rate of wage is likely to expand the aggregate demand for labour by for labour by not less than three percent" (Pigou, 1933). Or, to put the matter the other way, "If wages are pushed up above the point of marginal productivity, the decrease in employment would normally be from three to four times as great as the increase in hourly rates" (Pigou, 1933) so that the total incomes of the workers would be reduced correspondingly. In Pigou's view, Even if these figures are taken to represent only the elasticity of the demand for labour revealed in a given period of the past and not necessarily to forecast that of the future, they deserve the most serious consideration (Pigou 1933, 96).

7.2.1. Modification of Capitalism

Though Pigou was a supporter of capitalism, he believed that it required some modification. He held that a moderate form of socialism was needed to solve certain important is a purely psychological one since he emphasized on the demand for labour and rejected the role of investment. He held that cyclical changes take place on account of changes in the demand schedule of labour. The main force underlying this schedule is the expectations regarding prospective yield. In this way he rejected the investment element and emphasised the psychological factor.One of the key thinkers in twentieth century Development Studies was W.W. Rostow, an American economist and government official. Prior to Rostow, approaches to development had been based on the assumption that "modernization" was characterized by the Western world which were able to advance from the initial stages of underdevelopment. Accordingly, other countries should model themselves after the West, aspiring to a "modern" state of capitalism and a liberal democracy. Using these ideas, Rostow penned his classic *Stages of Economic Growth* in 1960, which presented five steps through which all countries must pass to become developed: 1) traditional society, 2) preconditions to take-off, 3) take-off, 4) drive to maturity, and 5) age of high mass consumption. The model asserted that all countries exist somewhere on this linear spectrum, and climb upward through each stage in the development process:

It is possible to identify all societies, in their economic dimensions, as lying within one of five categories: the traditional society, the preconditions for take-off, the take-off, the drive to maturity, and the age of high mass-consumption.

Stage 1–The Traditional Society

First, the traditional society. A traditional society is one whose structure is developed within limited production functions, based on pre-Newtonian science and technology, and on pre-Newtonian attitudes towards the physical world. Newton is here used as a symbol for that watershed in history when men came widely to believe that the external world was subject to a few knowable laws, and was systematically capable of productive manipulation. The conception of the traditional society is, however, in no sense static; and it would not exclude increases in output. Acreage could be expanded; some ad hoc technical innovations, often highly productive innovations, could be introduced in trade, industry and agriculture; productivity could rise with, for example, the improvement of irrigation works or the discovery and diffusion of a new crop. But the central fact about the traditional society was that a ceiling existed on the level of attainable output per head. This ceiling resulted from the fact that the potentialities which flow from modern science and technology were either not available or not regularly and systematically applied. Both in the longer past and in recent times the story of traditional societies was thus a story of endless change. The area and volume of trade within them and between them fluctuated, for example, with the degree of political and social turbulence, the efficiency of central rule, the upkeep of the roads. Population-and, within limits, the level of life-rose and fell not only with the sequence of the harvests, but with the incidence of war and of plague. Varying degrees of manufacture developed;

but, as in agriculture, the level of productivity was limited by the inaccessibility of modern science, its applications, and its frame of mind. Generally speaking, these societies, because of the limitation on productivity, had to devote a very high proportion of their resources to agriculture; and flowing from the agricultural system there was a hierarchical social structure, with relatively narrow scope-but some scope-for vertical mobility. Family and clan connexions played a large role in social organization. The value system of these societies was generally geared to what might be called a long-run fatalism; that is, the assumption that the range of possibilities open to one's grandchildren would be just about what it had been for one's grandparents.Although central political rule-in one form or another-often existed in traditional societies, transcending the relatively self-sufficient regions, the centre of gravity of political power generally lay in the regions, in the hands of those who owned or controlled the land. The landowner maintained fluctuating but usually profound influence over such central political power as existed, backed by its entourage of civil servants and soldiers, imbued with attitudes and controlled by interests transcending the regions.

Stage 2–The Preconditions for Take-off

The second stage of growth embraces societies in the process of transition; that is, the period when the preconditions for take-off are developed; for it takes time to transform a traditional society in the ways necessary for it to exploit the fruits of modern science, to fend off diminishing returns, and thus to enjoy the blessings and choices opened up by the march of compound interest. The preconditions for take-off were initially developed, in a clearly marked way, in Western Europe of the late seventeenth and early eighteenth centuries as the insights of modern science began to be translated into new production functions in both agriculture and industry, in a setting given dynamism by the lateral expansion of world markets and the international competition for them. But all that lies behind the break-up of the Middle Ages is relevant to the creation of the preconditions for take-off in Western Europe. Among the Western European states, Britain, favoured by geography, natural resources, trading possibilities, social and political structure, was the first to develop fully the preconditions for takeoff. The more general case in modern history, however, saw the stage of preconditions arise not endogenously but from some external intrusion by more advanced societies. These invasions-literal or figurative-shocked the traditional society and began or hastened its undoing; but they also set in motion ideas and sentiments which initiated the process by which a modern alternative to the traditional society was constructed out of the old culture.

The idea spreads not merely that economic progress is possible, hut that economic progress is a necessary condition for some other purpose, judged to be good: be it national dignity, private profit, the general welfare, or a better life for the children. Education, for some at least, broadens and changes to suit the needs of modern economic activity. New types of enterprising men come forward-in the private economy, in government, or both-willing to mobilize savings and to take risks in pursuit of profit or modernization. Banks and other institutions for mobilizing capital appear. Investment increases, notably in transport, communications, and in raw materials in which other nations may have an economic interest. The scope of commerce, internal and external, widens. And, here and there, modern manufacturing enterprise appears, using the new methods. But all this activity proceeds at a limited pace within an economy and a society still mainly characterized by traditional low-productivity methods, by the old social structure and values, and by the regionally based political institutions that developed in conjunction with them. In many recent cases, for example, the traditional society persisted side by side with modern economic activities, conducted for limited economic purposes by a colonial or quasi-colonial power. Although the period of transition-between the traditional society and the take-off-saw major changes in both the economy itself and in the balance of social values, a decisive feature was often political. Politically, the building of an effective centralized national state-on the basis of coalitions touched with a new nationalism, in opposition to the traditional landed regional interests, the colonial power, or both, was a decisive aspect of the preconditions period; and it was, almost universally, a necessary condition for take-off.

Stage 3–The Take-off

We come now to the great watershed in the life of modern societies: the third stage in this sequence, the take-off. The take-off is the interval when the old blocks and resistances to steady growth are finally overcome. The forces making for economic progress, which yielded limited bursts and enclaves of modern activity, expand and come to dominate the society. Growth becomes its normal condition. Compound interest becomes built, as it were, into its habits and institutional structure. The take-off awaited not only the build-up of social overhead capital and a surge of technological development in industry and agriculture, but also the emergence to political power of a group prepared to regard the modernization of the economy as serious, high-order political business. During the take-off, the rate of effective investment and savings may rise from, say, 5 % of the national income to 10% or more; although where heavy social overhead capital investment was required to create the technical preconditions for take-off the investment rate in the preconditions period could be higher than 5%. During the take-off new industries expand rapidly, yielding profits a large proportion of which are reinvested in new plant; and these new industries, in turn, stimulate, through their rapidly expanding requirement for factory workers, the services to support them, and for other manufactured goods, a further expansion in urban areas and in other modern industrial plants. The whole process of expansion in the modern sector yields an increase of income in the hands of those who not only save at high rates but place their savings at the disposal of those engaged in modern sector activities. The new class of entrepreneurs expands; and it directs the enlarging flows of investment in the private sector. The economy exploits hitherto unused natural resources and methods of production. New techniques spread in agriculture as well as industry, as agriculture is commercialized, and increasing numbers of farmers are prepared to accept the new methods and the deep changes they bring to ways of life. The revolutionary changes in agricultural productivity are an essential condition for successful take-off; for modernization of a society increases radically its bill for agricultural products. In a decade or two both the basic structure of the economy and the social and political structure of the society are transformed in such a way that a steady rate of growth can be, thereafter, regularly sustained.

Stage 4: The Drive to Maturity

After take-off there follows a long interval of sustained if fluctuating progress, as the now regularly growing economy drives to extend modern technology over the whole front of its economic activity. Some 10-20% of the national income is steadily invested, permitting output regularly to outstrip the increase in population. The make-up of the economy changes unceasingly as technique improves, new industries accelerate, and older industries level off. The economy finds its place in the international economy: goods formerly imported are produced at home; new import requirements develop, and new export commodities to match them. The society makes such terms as it will with the requirements of modern efficient production, balancing off the new against the older values and institutions, or revising the latter in such ways as to support rather than to retard the growth process. Some sixty years after take-off begins (say, forty years after the end of take-off) what may be called maturity is generally attained. The economy, focused during the take-off around a relatively narrow complex of industry and technology, has extended its range into more refined and technologically often more complex processes; for example, there may be a shift in focus from the coal, iron, and heavy engineering industries of the railway phase to machine-tools, chemicals, and electrical equipment. This, for example, was the transition through which Germany, Britain, France, and the United States had passed by the end of the nineteenth century or shortly thereafter. Formally, we can define maturity as the stage in which an economy demonstrates the capacity to move beyond the original industries which powered its take-off and to absorb and to apply efficiently over a very wide range of its resources-if not the whole range-the most advanced fruits of (then) modern technology. This is the stage in which an economy demonstrates that it has the technological and entrepreneurial skills to produce not everything, but anything that it chooses to produce. It may lack (like contemporary Sweden and Switzerland, for example) the raw materials or other supply conditions required to produce a given type of output economically; but its dependence is a matter of economic choice or political priority rather than a technological or institutional necessity. Historically, it would appear that something like sixty years was required to move a society from the beginning of take-off to maturity. Analytically the explanation for some such interval may lie in the powerful arithmetic of compound interest applied to the capital stock, combined with the broader consequences for a society's ability to absorb modern technology of three successive generations living under a regime where growth is the normal condition. But, clearly, no dogmatism is justified about the exact length of the interval from take-off to maturity.

Stage 5: The Age of High Mass-Consumption

We come now to the age of high mass-consumption, where, in time, the leading sectors shift towards durable consumers' goods and services: a phase from which Americans are beginning to emerge; whose not unequivocal joys Western Europe and Japan are beginning energetically to probe; and with which Soviet society is engaged in an uneasy flirtation. As societies achieved maturity in the twentieth century two things happened: real income per head rose to a point where a large number of persons gained a command over consumption which transcended basic food, shelter, and clothing; and the structure of the working force changed in ways which increased not only the proportion of urban to total population, but also the proportion of the population working in offices or in skilled factory jobs-aware of and anxious to acquire the consumption fruits of a mature economy. In addition to these economic changes, the society ceased to accept the further extension of modern technology as an overriding objective. It is in this post-maturity stage, for example, that, through the political process, Western societies have chosen to allocate increased resources to social welfare and security. The emergence of the welfare state is one manifestation of a society's moving beyond technical maturity; but it is also at this stage that resources tend increasingly to be directed to the production of consumers' durables and to the diffusion of services on a mass basis, if consumers' sovereignty reigns. The sewing-machine, the bicycle, and then the various electric-powered household gadgets were gradually diffused. Historically, however, the decisive element has been the cheap mass automobile with its quite revolutionary effects-social as well as economic-on the life and expectations of society.

7.3. Dynamic Theory of Production

These stages are not merely descriptive. They are not merely a way of generalizing certain factual observations about the sequence of development of modern societies. They have an inner logic and continuity. They have an analytic bone-structure, rooted in a dynamic theory of production. The classical theory of production is formulated under essentially static assumptions which freeze-or permit only once-over change- in the variables most relevant to the process of economic growth. As modern economists have sought to merge classical production theory with Keynesian income analysis they have introduced the dynamic variables: population, technology, entrepreneurship etc. But they have tended to do so in forms so rigid and general that their models cannot grip the essential phenomena of growth, as they appear to an economic historian. We require a dynamic theory of production which isolates not only the distribution of income between consumption, saving, and investment (and the balance of production between consumers and capital goods) but which focuses directly and in some detail on the composition of investment and on developments within particular sectors of the economy. The argument that follows is based on such a flexible, disaggregated theory of production. When the conventional limits on the theory of production are widened, it is possible to define theoretical equilibrium positions not only for output, investment, and consumption as a whole, but for each sector of the economy. Within the framework set by forces determining the total level of output, sectoral optimum positions are determined on the side of demand, by the levels of income and of population, and by the character of tastes; on the side of supply, by the state of technology and the quality of entrepreneurship, as the latter determines the proportion of technically available and potentially profitable innovations actually incorporated in the capital stock.

William Jevons was one of three men to simultaneously advance the so-called marginal revolution. Working in complete independence of one another -Jevons in Manchester, England; LeonWalras in Lausanne, Switzerland; and CarlMenger in Vienna - each scholar developed the theory of marginal utility to understand and explain consumer behaviour. The theory held that the utility (value) of each additional unit of a commodity -the marginal utility- is less and less to the consumer. When you are thirsty, for example, you get great utility from a glass of water. Once your thirst is quenched, the second and third glasses are less and less appealing. Feeling waterlogged, you will eventually refuse water altogether. "Value," said Jevons, "depends entirely upon utility."This statement marked a significant departure from the classical theory of value, which stated that value derived from the labour used to produce a product or from the cost of production more generally. Thus began the neoclassical school, which is still the dominant one in economics today.

Jevons went on to define the "equation of exchange," which shows that for a consumer to be maximizing his or her utility, the ratio of the marginal utility of each item consumed to its price must be equal. If it is not, then he or she can, with a given income, reallocate consumption and get more utility. Take, for example, a consumer whose marginal utility from oranges is 10 "utils," and from cookies 4 utils, when oranges and cookies are both priced at $.50 each. The consumer's ratio of marginal utility to price for oranges is 10/$.50, or 20, and for cookies is 4/$.50, or 8. Jevons would have said (and modern economists would agree) that this does not satisfy the equation of exchange, and therefore the consumer will change purchases. Specifically, the consumer could increase utility by spending $.50 less on cookies and using the money to buy oranges. He would lose 4 utils on the cookies, but gain 10 on the oranges, for a net gain of 6 utils. He will have this incentive to reallocate purchases until the equation of exchange holds (*i.e.*, until the marginal utility of oranges falls and the marginal utility of cookies rises to a point where, as a ratio to their prices, they are equal). Jevons put much less thought into the production side of economics. It is ironic, therefore, that he became famous in Britain for his book The Coal Question, in which he wrote that Britain's industrial vitality depended on coal and, therefore, would decline as that resource was exhausted. As coal reserves ran out, he wrote, the price of coal would rise. This would make it feasible for producers to extract coal from poorer or deeper seams. He also argued that America would rise to become an industrial superpower. Although his forecast was right for both Britain and America, and he was right about the incentive to mine more costly seams, he was almost surely wrong that the main factor was the cost of coal. Jevons failed to appreciate the fact that as the price of an energy source rises, entrepreneurs have a strong incentive to invent, develop, and produce alternate sources. In particular, he did not anticipate oil or natural gas. Also, he did not take account of the incentive, as the price of coal rose, to use it more efficiently or to develop technology that brought down the cost of discovering and mining.

7.3.1. Main Works and Ideas

At the age of 21, he began to develop his own structure of economics by using mathematics. A few years later, before he turned 25, Jevons discovered the utility theory or the "true theory of economy," which was considered to be the beginning point of his analysis in economics, and this led to his development of theory of exchange. He published this theory along with expositions of labour, rent, and capital. Due to his earlier rigorous training in natural science, economics in Jevons's views demanded a mathematical treatment. He argued that the subject deals with quantities, so it has to consistently employ differential calculus (with its infinitesimally tiny quantities) and graphic representation. Furthermore, he expected that the mathematical treatment would prove to be successful tool as in physics and chemistry. Jevons' major contribution to economic theory was in the publication in 1871 of The Theory of Political Economy.

(A) Jevon's Utility Theory

Jevons's utility theory was based on Bentham's approach. In addition, much from Bentham, he described economics as a "calculus of pleasure and pain." He acknowledged that feelings of human are matter, so the quantitative effects of the feelings should be measured and compared. To define utility, Jevons described that not just intrinsic quality of things that count, but also "a circumstance of things arising out of their relation to man's requirements". That means a certain quantity of the same commodity produces different utilities to different men. A commodity will be needed up to a certain quantity. Beyond that quantity, a commodity will produce less and less utility down to zero or even negative. Along with these observations, Jevons attempted to differentiate between total utility and thing later known as "marginal utility," which introduced by Wicksteed and Marshall. At the time, he called marginal utility as "final degree of utility," which defined as "the degree of utility of the last addition, or the next possible addition of a very small, or infinitely small, quantity to the existing stock". This concept in the language of differential calculus is actually the differential coefficient of utility with respect to quantity. Jevons's utility theory reached the highest point where he described maximum total utility of a commodity that has various applications is reached when the final degrees of utility are equal in every use. Along with his analysis, he used equations and diagrams (quantities of the good on the x-axis and utility on the y-axis) and the area under the curve to represent total utility. Jevons's utilitarianism was still not a perfect concept. According to Jevons, the calculus of utility and the higher calculus of moral right and wrong were different. He mentioned that the latter was superior to the former except in matters of moral indifference. Jevonswas considered to be an advocate of the ordinal measurement of utility. He described that the possibility of measuring the total amount of pleasure or comparing one pleasure as an exact multiple of another was seldom or never. His theory had no need of the comparison of total quantities of utility instead it was based on the comparison of equal small increments. Moreover, Jevons rejected the idea of interpersonal comparisons of utility, which he considered not possible. Jevons developed theory of exchange based upon his utility theory. The prerequisites of this theory are such that:

I. Greater than two participants dealing with more than two commodities

II. Perfect knowledge in stocks, intentions of the exchangers, and the ratio of exchange Market idea of his covered the situations from bilateral monopoly to competition. The theory suggests that the ratio of exchange of any two goods "will be the reciprocal of the ratios of the marginal utilities of the quantities of goods available for consumption after completion of the exchange". Basically, people will continued to trade as long as marginal utility of the acquired commodity is greater than marginal utility that losses by the surrender of the other commodity. So, due to Jevons's concept of diminishing marginal utility as described previously, the trading process will stop when the marginal utilities of both goods (the good acquired and the good surrendered) are equivalent. In connection with Jevons's market concept, the perception of "trading bodies" on the buying and selling sides, which comprised of single individuals or groups of individuals, was also introduced. Jevons's also introduced the law of indifference. According to this law, in the same market "all portions of a homogeneous good must be exchanged in the same ratio, any portion being indifferently usable in place of an equal portion". The foundation of Jevons's law of indifference does still exist in modern economic theory – one price in one market. However, he did not successfully apply the equation of exchange to the competitive market because the solution requires a rather complex methodology. But this attempt was later achieved by Walras and Marshall.

(B) Jevon's Analysis of Real Cost

Even though he did not fully develop demand curves from utility curves and money cost from supply curves, he successfully achieved the development of the theory of real cost. He represented real cost in terms of feelings of disutility or pain. This theory led to the analysis of the determinants of the supply of labour. He described that at first workers would think of work as tedious and later work became pleasurable (because they saw an opportunity to increase their total utility by working) until a certain number of work hours beyond that work would become tedious once again. Workers would stop working once his marginal utility of his product or of his wages equalled to marginal disutility of effort. Amazingly, this great concept had small contribution on modern economic thought because its unrealistic assumption that individual worker can choose number of his working hours.

(C) The Ending of the Labour Theory of Value

The labour theory of value became obsolete after the introduction of Jevons's theory of utility. Jevons pointed out that the labour theory of value was not useful in many occasions. It only meant for the case of commodities that could be remanufactured by labour. In many cases, market prices of goods did not represent the quantities of labour used to produce them.

Jevons replaced labour by utility and scarcity as determinants of value. He also interpreted exchange value and use value in terms of marginal utility and total utility, respectively. In addition, he emphasized that some factors from the demand side could affect value instead of supply and cost. So, the value derived from the state of mind or a "distinct feeling" rather than the conditions of the external world.

He refused to use labour as a direct measurement of value, but he accepted that the value of a good might be implicitly determined by labour. He explained that the quantity of labour used in the production of a good can affect its supply, and supply will in turn influence on the degree of utility of the good, which determines the ratio of exchange. The following words are Jevon's expression of this chain of reasoning: "Cost of production determines supply; Supply determines final degree of utility; Final degree of utility determines value". However, this type of analysis contains an error. Instead of explaining a simultaneous relationship of mutual interdependence, it points out a time sequence with cause-and-effect relationships. Walras's work has mentioned this fact and served as a foundation of modern economic analysis.

(D) Jevon's Empirical Studies

Jevons was considered to be an exponent of pure mathematical economics rather than an econometrician. His empirical studies using knowledge of econometrics began in the early 1860s. His book named The Coal Question (1865) contained a combination of theoretical and empirical work. In this book, Jevons made a systematic statement of a "natural law of social growth." He explained that the significance of coal began to out pace the importance of corn in the British economy. He also included his observations in this book suggesting that the demand for coal would grow at a geometric rate and it might not be able to sustain in the long-run. Many of Jevons's empirical studies were collected in a volume titled Investigations in Currency and Finance (1884). The main focus of these studies is the issues of economic fluctuations. His approach used in the earlier studies was mainly the methods of natural science. He discovered three major fluctuations that determine the up and down of the economy – seasonal movements, business cycles, and secular trends. Later in his life, he discovered the link between the business cycle and ten year variations of the sunspots. Later he extended his contribution to index-number theory and innovations in graphic presentation, including ratio charts. Born in Liverpool, England, Jevons studied chemistry and botany at University College, London. Because of the bankruptcy of his father's business in 1847, Jevons left school to take up the position of assayer at the Mint in Sydney, Australia. He remained there five years, resuming his studies at University College on his return to England. He was later appointed to the post of chair in political economy at his alma mater and retired from there in 1880. Two years later, with a number of unfinished books in process, Jevons drowned while swimming. He was forty-six.

8

Alfred Marshall: His Importance in Neoclassical Economics

8.1. Alfred Marshall: His Importance in Neoclassical Economics

Alfred Marshall was the dominant figure in British economics (itself dominant in world economics) from about 1890 until his death in 1924. His specialty was MICROECONOMICS–the study of individual markets and industries, as opposed to the study of the whole economy. In his most important book, Principles of Economics, Marshall emphasized that the price and output of a good are determined by both supply and demand: the two curves are like scissor blades that intersect at equilibrium. Modern economists trying to understand why the price of a good change, still start by looking for factors that may have shifted demand or supply, an approach developed by Marshall. Marshall grew up in the London suburb of Clapham, being educated at the Merchant Taylor's School where he showed academic promise and a particular aptitude for mathematics. Eschewing the more obvious path of a closed scholarship to Oxford and a classical education, he entered St John's College, Cambridge, in 1862 on an open exhibition. There he read for the Mathematical Tripos, Cambridge University's most prestigious degree competition, emerging in 1865 in the exalted position of Second Wrangler, bettered only by the future Lord Rayleigh. This success ensured Marshall's election to a Fellowship at St John's. Supplementing his stipend by some mathematical coaching, and abandoning - doubtless because of a loss of religious conviction -half-formed earlier intentions of a clerical career, he became engrossed in the study of the philosophical foundations and moral basis for human behavior and social organization. In 1868 he became a College Lecturer in Moral Sciences at St John's, specializing in teaching political economy. By about 1870 he seems to have committed his career to developing this subject and helping to transform it into a new science of economics.

Marshall had no great profundity as a philosopher of science and had little patience with metaphysics. His discussions of methodology largely reflect the philosophical presuppositions of his day. His method was in the general deductive tradition of Ricardo, John Stuart Mill and Cairnes. But he sought to emphasize the relativity of particular theories, as contrasted with the universality of the general theoretical frame work. And he was anxious to choose his assumptions with close regard to the facts of the case. Marshall's method was described by J.N. Keynes as "deductive political economy guided by observation". To Marshall also goes credit for the concept of price elasticity of demand, which quantifies buyers' sensitivity, to price The concept of consumer surplus is another of Marshall's contributions. He noted that the price is typically the same for each unit of a commodity that a consumer buys, but the value to the consumer of each additional unit declines. A consumer will buy units up to the point where the marginal value equals the price. Therefore, on all units previous to the last one, the consumer reaps a benefit by paying less than the value of the good to himself. The size of the benefit equals the difference between the consumer's value of all these units and the amount paid for the units. This difference is called the consumer surplus, for the surplus value or utility enjoyed by consumers. Marshall also introduced the concept of producer surplus, the amount the producer is actually paid minus the amount that he would willingly accept. Marshall used these concepts to measure the changes in wellbeing from government policies such as taxation. Although economists have refined the measures since Marshall's time, his basic approach to what is now called welfare economics still stands. Wanting to understand how markets adjust to changes in supply or demand over time, Marshall introduced the idea of three periods. First is the market period, the amount of time for which the stock of a commodity is fixed. Second, the short period is the time in which the supply can be increased by adding labour and other inputs but not by adding capital (Marshall's term was "appliances"). Third, the long period is the amount of time taken for capital ("appliances") to be increased. To make economics dynamic rather than static, Marshall used the tools of classical mechanics, including the concept of optimization. With these tools he, like neoclassical economists who have followed in his footsteps, took as givens technology, market institutions, and people's preferences. But Marshall was not satisfied with his approach. He once wrote that "the Mecca of the economist lies in economic biology rather than in economic dynamics." In other words, Marshall was arguing that the economy is an evolutionary process in which technology, market institutions, and people's preferences evolve along with people's behavior. Marshall rarely attempted a statement or took a position without expressing countless qualifications, exceptions, and footnotes. He showed himself to be an astute mathematician–he studied math at St. John's College, Cambridge–but limited his quantitative expressions so that he might appeal to the layman. Marshall was born into a middle-class family in London and raised to enter the clergy. He defied his parents' wishes and instead became an academic in mathematics and economics.

8.2. Production and Long-Period Competitive Supply

In deriving the long-period supply curve of a commodity, Marshall envisages production as organized by firms, typically family businesses. Each firm strives to minimize its production costs, substituting one productive factor or production method for another according to the "Principle of Substitution." In its simpler forms this involves marginalist adjustment to bring relative marginal value products into line with relative marginal costs. Marshall's firms do not have costless access to a common production function, but must grope and experiment their way to cost-reducing modifications. The long period supply curve is defined for a given state of *general* scientific and technical knowledge. The long-period supply price per unit of output at which such an industry can supply any quantity of output must just cover the cost of maintaining that level of output indefinitely. That is, it must just suffice to pay all the inputs (including management) needed to produce that level of output in a cost-minimizing way at rates which will just ensure that the requisite input quantities will continue to be forthcoming. In the case of skilled workers, in particular, the rate must just suffice to induce parents to apprentice new workers to the industry at a rate exactly offsetting the attrition through retirement,etc. Similarly, the return to fixed capital must just suffice to induce replacement of the existing stock of fixed assets, while the return to management must keep up the necessary replacement flow of managers. On the other hand, the return to land services must just suffice to prevent these services from migrating elsewhere, replacement not being necessary. As the level of industry output being considered is increased, the supply price will probably rise, mainly because of the need to pay a higher return to land so as to attract a greater supply from other uses, but perhaps also because of the need to pay more for rare natural talents which, like land, must be attracted in greater quantity from other uses, not being capable of replication through education and training. Such a tendency for long-period supply price to rise with output may be mitigated though seldom eliminated by substitution against inputs whose supply price is rising and by possible external economies which increase each firm's efficiency by influences which depend, not on its own output, but on the entire industry's output. A tendency for supply price to rise with output will imply that inframarginal units of those inputs whose supply prices are rising receive rents, since all units will be remunerated at the rate necessary to induce continuing supply of the marginal unit.

(A) Representative Firm

It is here that Marshall's "representative firm" enters the picture. It is best regarded as a parable which avoids the need to consider the entire distribution of firms. By definition, the long-period supply price of any level of industry output is the average cost of the representative firm at that level of output. Industry-level magnitudes may then be regarded as if they were generated by a fixed number of unchanging representative firms rather than by the actual heterogeneous body of ever-changing firms– that is, the manufacturing case may be treated as if it were an agricultural case. Such arguments add nothing conceptually and are prone to confuse, although it might be noted that Marshall believed an acute well-informed observer could select an actual firm which was close to being representative in this sense. The average cost and size of the representative firm will change as *industry* output changes. There are two main reasons for this. A larger industry output is likely to generate more *external* economies, lowering the costs of every firm. But more importantly, the larger is industry demand the easier it will be for a new firm to build up a market, and so the larger the size to which firms will grow before they begin to decline. This will bring about greater access on average to unexhausted *internal* economies of scale, again leading to lower costs on average. For both these reasons, long-period supply price is likely to decline as a larger industry output is considered, even though the opportunity cost of obtaining greater supplies of land services and rare natural talents may rise. Again, the particular expenses curve may be used to display the surpluses or rents accruing to such scarce factors at any given level of industry output, but the relationship of this family of curves to the long-period supply curve is tenuous and complex. Rent obviously cannot be represented by a "triangle" above the supply curve when the latter is falling. The conception of competition in Marshall's manufacturing case is much closer to later ideas of imperfect or monopolistic competition than to modern notions of perfect competition. Products are differentiated and firms are not price takers, but face at any time downward-sloping demand curves in their special markets. Even if the difficulties of rapidly building up a firm's internal organization can be overcome, the resulting enlarged output cannot be sold at a price covering cost–even granted substantial scale economies in production–without going through the slow process of building up a clientele and shifting the firm's particular demand curve. The time this takes is assumed to be considerable relative to the duration of the firm's initial vitality. But in some cases the difficulties of rapid expansion may be overcome. They may not have been very severe, as when different firms' products are highly substitutable, or the firm's founder may have unusual genius. In such cases the industry will pass into a monopoly or be dominated be a few, strategically-interacting firms, or "conditional monopolies" as Marshall termed them.Marshall's reconciliation of persisting competition with increasing returns and falling supply price is complex and problematic, but it does not depend in any essential way on scale economies being external to the firm. The concept of external economies is one of his significant contributions, although his treatment of it can hardly be called pellucid. But it was added more for verisimilitude than because it was theoretically essential to the structure of his theory.

(B) Price Determination and Period Analysis

The long-period supply curve for any good indicates for each market quantity the least price at which that quantity will continue indefinitely to be supplied. The equilibrium price and quantity (long period) are determined by the intersection of this supply curve with the negatively sloped market demand curve, indicating the highest uniform price at which any total quantity can be sold. In an agricultural case, equilibrium will be unique as the supply curve slopes positively. But in a manufacturing case, the supply curve, as well as the demand curve, will have negative slope, so that multiple equilibria can occur. Equilibrium is adjudged locally stable if demand price is above (below) supply price at a quantity just below (above) the equilibrium quantity. The intuitive justification for this is that the actual price of any available quantity is determined by the demand price, while quantity produced tends to increase (through both expansion of existing firms and rapid entry of new firms) whenever an excess of market price over supply price promises high profits, while it tends to decrease in the opposite case. Marshall emphasized three broad cases. Temporary or market equilibrium analysis proceeded on the assumption of a fixed stock of output already available or in the pipeline. Short-period–normal equilibrium analysis permitted output to be varied but not the stock of productive "appliances" available to produce that output. "Appliances" must be taken here to cover skilled labour and business organization as well as fixed capital assets, so that the existing set of firms is to be taken as given. Finally, long-period–normal equilibrium, which has already been considered, allows the stock of appliances to be freely varied, as well as the level of output. In this case equilibrium incorporates the conditions necessary for inducing a replacement flow of each kind of appliance, including a replacement flow of new firms in the manufacturing case.

Temporary equilibrium for a perishable commodity is simply a matter of selling off the existing stock. Marshall recognizes the possibility of "false trading"–sales at a non-equilibrium price–but argues that (a) this will not affect the eventual price if the marginal utility of money is constant, and (b) price will quickly settle close to that uniform price which would just clear the market if used in all transactions. With a storable good there is the further speculative possibility of holding back supply for future sale, and this gives expected future cost of production an indirect role in influencing current market price. Cost of production already incurred is an irrelevant bygone, however. In short-period–normal equilibrium, output is adapted to demand within the constraints set by the fixed supply of available "appliances." High demand will raise equilibrium output, but only within the limits possible by working existing appliances more intensively or pulling in versatile unspecialized labour and land from elsewhere. Low demand will lead to low utilization of appliances, perhaps idleness of some, and migration of unspecialized inputs to elsewhere. In the agricultural case a firm will change output until marginal prime or variable cost equals market price. In the manufacturing case, a fear of spoiling the future market or invoking retaliation from competitors tends to make a firm's output more responsive to variation in market price, and hence to make market price less responsive to demand shifts. Otherwise, the two cases are similar, both involving a fixed population of firms and a rising supply curve. The return received by an appliance will often exceed the minimum necessary to induce its operation at the chosen intensity (its prime cost) and this excess is a "quasi rent." To the extent that land and rare natural talents are immobile in the short period, or less mobile in the short period than the long, then their returns too will often have a quasi-rent element. Otherwise, they will receive only differential rents, though often at rates differing from their long-period values. It should be stressed that the concepts of quasi rent and differential rent are relative to a specific use. The prime cost necessary to retain an input in this use may itself include rent or quasi rent when viewed in the context of a more inclusive set of alternative uses. Thus, from the viewpoint of all possible uses in the economy, the return to any factor in fixed supply is entirely a rent or quasi rent (the latter if fixity is only short-period).

Marshall's period analysis, and more generally his partial-equilibrium approach to price determination, was designed in large part as a usable tool for the analysis of concrete issues. Its longevity amply testifies to its usefulness in this respect. But it was also meant to serve the more doctrinal purpose of clarifying the respective roles utility and cost of production play in determining value. The aim was to show that the greater the scope for supply adjustment permitted in the definition of equilibrium the more dominant the supply[-]side influence on price becomes. This doctrinal goal helps to account for the rather heavy weight given to long-period analysis in the *Principles*. For, as Marshall recognized, its value as a tool of applied analysis is seriously qualified by the fact that "violence is required for keeping broad forces in the pound of Caeteris Paribus during, say, a whole generation, on the ground that they have only an indirect bearing on the question in hand". That is, there is no good ground for assuming that background forces such as technology and tastes will remain constant for the length of time required for long-period equilibrium to be practically relevant. For concrete analysis of problems of such long duration it will often be necessary to transcend the period analysis, with its reliance on statical equilibrium, and undertake directly an analysis of secular change, of which Book VI, Ch. XII of the *Principles* on the "General Influence of Economic Progress" offers the main example, but not a very impressive one.

(C) Normal Value and Normal Profit

Implicit in the preceding discussion are Marshall's conceptions of normal value and normal profit. Normal value is defined as the value which would result "if the economic conditions under view had time to work out undisturbed their full effect". It is contrasted with market value, which is "the actual value at any time". Normal value is hypothetical, resting on a *ceteris paribus* condition, its role being to indicate underlying tendencies. The normal value of a commodity may approximate its average value over periods sufficiently long for the "fitful and irregular causes" which dominate market value to cancel out, but this should not be presupposed automatically outside a hypothetical stationary state. The distinction between normal and market value is closely related to the distinction between natural and market value found in the work of Smith and the classical economists. In 1879 Marshall had identified normal value with "the results which competition would bring about in the long run", but in the *Principles* he switched to the view that "Normal does not mean Competitive" and admitted any kind of regular influence so long as it was sufficiently persistent. The economic forces hypothetically permitted to achieve full mutual accommodation could now be chosen appropriately for each case. In particular, the distinction between short-period and long-period normal ("subnormal" and "truenormal" in earlier editions) was emphasized. Profit was viewed by Marshall as the residual income accruing to a firm's owner, a return to the investment of his own capital and to the pains he suffers in exercising his "business power" in planning, supervision and control. *Normal* profit is essentially an opportunity cost, the minimum return necessary to secure the owners' inputs to their current use, or rather to accomplish this for an owner of normal ability.

Marshall presumes that there is a large and elastic supply of versatile actual or potential owner managers of normal ability. In long-period equilibrium each of these must just receive the same normal rates of return on his investment and exercise of business power whatever his line of business. These common rates of normal return are simultaneously determined, along with the normal returns to other kinds of effort and abstinence, by Marshall's macroeconomic theory of the long-period determination of factor incomes. Although profits are a residual, rather than a contractually agreed amount like other incomes, the difference is immaterial in long-period equilibrium. In particular, a long-period equilibrium analogy between ordinary wages and the normal earnings of business power is a necessary element in the costs which underlie the long-period normal supply curve, but actual profit is a quasi-rent or surplus for shorter periods.

Normal profits are a return to "business power in command of capital" and compensate for three distinct elements: "the supply of capital, the supply of the business power to manage it, and the supply of the organization by which the two are brought together and made effective for production". The combined compensation of the latter two components comprises "gross earnings of management," the return to the second component being "net earnings of management." The normal return to the first element is imputed at the market interest rate on default-free loans, and that to the second component at the rate paid to hired managers performing comparable tasks. The residual third element, the return to "organization," is most straightforwardly interpreted as an extra return on owned capital equivalent to the premium for default risk, or "personal risk," which would have to be paid on borrowed capital. In the manufacturing case, the annual level of normal profit for each firm in an industry must be interpreted as the annualized equivalent of the expected stream of returns just sufficient to induce an individual of normal ability to found a firm in the industry rather than divert his energies and capital elsewhere. Normal ability here is defined relative to other potential founders of firms, a group already exceptional relative to the population as a whole. By construction, such normal profits must be earned by the representative firm.

(D) Welfare Economics

To serve as a tool of welfare economics, monetary measures of changes in consumer surplus, producer surplus and rent must be aggregated over individuals, but how are the resulting sums to be interpreted? Marshall's very limited and proximate attempts at formal welfare arguments are carried out within a utilitarian framework, for which the goal is maximizing aggregate utility. He implies that interpersonal utility comparisons are possible in principle and that utility functions will be similar for all members of any group that is homogeneous in terms of mental, physical and social attributes. Within such a group, the marginal utility of money will be the same for two individuals having the same income, and lower for the richer of two individuals having different incomes, assuming in each case that both individuals face the same trading opportunities. A postulated change (*e.g.* a government action) will impose gains and losses on individuals which can be measured and aggregated in money-equivalent terms, but how can these measures be translated into statements about aggregate gains and losses of utility? Marshall emphasizes two special cases. If the gains are distributed over groups, and over income classes within each group, in exactly the same proportions as the losses are distributed, then aggregate utility gain will stand in the same proportion to aggregate monetary gain as aggregate utility loss stands to aggregate monetary loss. Even without knowing this proportion, the aggregate net monetary gain or loss will serve as an index of the aggregate utility gain or loss (although it can only rank alternatives having the same *relative* distributions of monetary benefits and costs as the case in question).

Alternatively, if money gains and losses are distributed similarly over groups, but within each group the gains accrue to individuals of lower income than those bearing the costs, then there must be an aggregate net utility gain even if the aggregate net monetary gain is zero-a warrant for certain redistributive policies. Marshall believed that these special cases were of quite wide applicability. In other cases, he saw that careful judgemental assessments of the marginal utility of money to the various injured and benefited groups would be necessary, assessments which could be used to transform monetary gains and losses into utility measures. But he gave little indication as to how these assessments might be obtained in practice. Marshall's best known and most successful foray into formal welfare analysis was his proof that total welfare might be increased by using the proceeds of a tax on an "agricultural" industry to subsidize a "manufacturing" industry. All comparisons involved long-period equilibria and relied on the validity of aggregated money-equivalent measures of gains and losses. He demonstrated that the gain in consumer surplus in the expanded decreasing-cost manufacturing industry might exceed the combined loss in consumer surplus and producer rents in the contracted increasing-cost agricultural industry. No formal account was taken of the possible gain in producer rent in the manufacturing industry as this merely made the argument *a fortiori*. The crucial point in this argument, as Marshall recognized, is that producers are not harmed by "a fall in price which results from improvements in industrial organization" (1920, p. 472). It is immaterial whether the improved organization of the enlarged manufacturing industry is due to external economies or to internal economies resulting from an increase in the size of the representative firm. Contrary to much subsequent opinion, Marshall's tax-subsidy argument is not necessarily dependent upon external economies. Another significant, but overlooked, welfare analysis provided by Marshall was that of a monopolistic public enterprise in a situation where taxation involves an excess burden (1920, pp. 487– 93, 857–8). Marshall proposes the goal of "compromise benefit" which effectively sums consumer surplus and monopoly revenue, but the latter multiplied by the marginal cost of raising a unit of government revenue from other sources. Maximization of compromise benefit leads to the setting of what has come to be termed a Ramsey price. In the absence of an excess burden, when the marginal cost of extra government revenue is unity, this reduces to marginal cost pricing. The two examples of welfare analysis just described proceed within a partial equilibrium framework, treating each industry as negligible compared to the entire economy and regarding the marginal utility of money as approximately constant to each individual. The gains or losses to producers need only take account of the narrow differential advantages obtained by operating in the industry in question rather than in any other use. Marshall's rather fragmentary remarks on optimal tax systems, income redistribution, and the "doctrine of maximum satisfaction" cannot be restricted in this way, and so raises serious unresolved analytical difficulties. On the other hand, his tax-subsidy argument was offered as a valid counterexample to arguments that competition must lead to a social optimum, or that optimal indirect tax systems must involve uniform tax rates. It must also be borne in mind that utilitarian welfare economics was for Marshall only a proximate step towards a more evolutionary analysis of modes of improving *the physical quality and the values and activities of mankind.*

(E) Inter-Related Markets and Distribution Theory

Marshall was anxious to emphasize the interdependence of markets and introduced his treatment of joint and composite demand and supply largely for this purpose. A group of products is jointly supplied or demanded if its members are all the outputs of or inputs into a single activity. A product is compositely supplied or demanded if it is produced or consumed by several activities. Marshall's formal treatment of joint demand and supply proceeded on the general assumption of fixed proportions, as did the related analysis of the derived demand for any one of several jointly-demanded inputs. The derived demand curve for an input is effectively constructed on the assumption that, as the price of the input in question is arbitrarily varied, prices of the remaining inputs and of the output of the activity in which they are used always adjust to keep demand equal to supply. The prime example of joint demand is the demand for productive inputs, and Marshall's analysis of market interdependence was carried through more fully in this specific connection, the role of substitution among inputs receiving full acknowledgement. The principle of substitution ensured that input usage was adjusted by firms to equate marginal value product, taken normally as the value of the marginal product (or "net product"), to the unit price of the input. Interdependence among input markets was further highlighted in the analysis of the competition of several industries for an input which is in temporarily or permanently fixed overall supply. A peculiarity of this last analysis was the insistence on excluding from the marginal cost of any industry the cost of bidding such fixed resources away from other uses. This is a perfectly legitimate application of the general envelope theorem: providing resource use is optimally adjusted, the marginal cost of increasing output will be the same whatever input or subgroup of inputs is increased. But Marshall's insistence on asymmetry where there is really symmetry can only be accounted for by his desire to legitimate, and extend to quasi rent, the classical doctrine that rent is price determined rather than price determining. Marshall's vision of market interdependence culminates in his treatment of distribution, where he seeks to bring out the extents to which the interests of different factors are consilient or conflicting. Distribution is determined by the interaction of the demands and supplies for the various inputs, the demands being essentially joint demands. Marginal productivity is a theory of input demand, not a complete theory of distribution, because the supplies of the various inputs cannot be viewed as fixed, at least in the long period. Indeed, in the long period the dominant influences on the prices of factors other than land are exerted by their supply conditions, the costs which have to be met in order for various kinds of labour and capital to just continue to be replaced in their existing uses and quantities. From an overall view "The net aggregate of all the commodities produced is itself the true source from which flow the demand prices for all these commodities, and therefore for the agents of production used in making them" (1920, p. 536). This aggregate, "the national dividend," is distributed among the factors of production. All have a common interest in increasing the size of the pie to be shared, but each has a selfish interest in restrictive practices which increase its own share even if they reduce the pie slightly. A prime question of social policy for Marshall is how these divergent incentives can be reconciled: how combined action by various groups, such as unions, can be prevented from assuming forms which, while perhaps individually beneficial to any one group in isolation, are certainly mutually harmful if undertaken by all.

Marshall here enters into macroeconomic forms of argument, and it is indeed true that he did toy with the formal specification of macroeconomic models of growth and distribution (Whitaker, 1975, Vol. II, pp. 305–16). But, with this exception, it should be emphasized that his treatment of market interdependence fell far short of a full theory of general equilibrium on Walrasian lines. Even when formalizing market interdependence in the mathematical appendix to the *Principles* (1920, pp. 846–56), he simply treated the demand or supply of each commodity as a function of nothing but the price of the commodity itself. The links between the generation of income in factor markets and the expenditure of that income in product markets were left quite vague. Again, it must be recalled that the development of comprehensive fully articulated equilibrium theories was not Marshall's aim.

(F) Monopoly and Combination

Marshall's analysis of price and output determination by a profit-maximizing monopolist, and of the effects of taxing such a monopolist, followed the lead of cournot. the concept of marginal revenue was implicitly in the mathematical statement, but marshall's chosen vehicle was geometrical. curves of average revenue and cost, and of their difference, average net revenue, y, (all functions of the quantity sold, x) were superimposed on a grid of iso-profit hyperbolae of form xy = constant. profit was maximized when the average net revenue curve touched the highest such iso-profit curve. Weighting consumer surplus into the maximand, as well as net revenue, gave rise to the welfare analysis of "compromise benefit" already mentioned. monopoly analysis was applied to trades unions, using the concept of the derived demand for an input, in which context marshall laid down his four rules for inelastic derived demand. these were that the input should have no good substitutes, that the product it helps make should be inelastically demanded, that the input should account for only a small part of production costs, and that cooperating inputs should be inelastically supplied. a union controlling a labour input for which derived demand is inelastic can certainly raise wages–not only the wage rate but the total wages received–although at the price of unemployment of some members. whether such a monopolistic restriction can be sustained for long is more doubtful, as there will be pressures both to enter the union and to evade its grasp by the relocation or reorganization of production. a more problematic question was whether "labour's disadvantage in bargaining" meant that combined action by workers could raise wages, even without any restriction of labour supply. Marshall believed that it did, but emphasized that the result might be less capital accumulation by non-workers, an outcome which could harm workers eventually. The extremes of monopoly and competition were both covered by the theory of normal value, even though the competition might be more akin to later concepts of imperfect or monopolistic competition than to any ideal form of perfect competition. But "normal action falls into the background, when trusts are striving for the mastery of a large market" (1920, p. xiv). the incidents, tactics and alliances of oligopolistic conflict defied reduction to a simple general theory. They were to have been considered in the uncompleted second volume of the *principles* and were to some extent covered by *industry and trade*. the latter's treatment of entry-limiting behaviour by a "conditional monopolist," who dominates the market but does not control entry, is of considerable interest in the light of much recent work on this class of problems.

9
Monetary Theory

9.1. Monetary Theory

Marshall was in full command of previous British discussions of monetary issues, but not himself a major contributor to the development of monetary theory. His evidence before Royal Commissions in 1887 and 1899 showed an impressive mastery of monetary analysis, both domestic and international, and was minutely examined by successive generations of Cambridge students, serving for many years virtually as a text book. But it was not until 1923, with the appearance of *Money, Credit and Commerce,* that Marshall put forward his monetary views in a systematic way. By then these had not the novelty, nor the vigour, to advance contemporary discussion. Marshall's most important contribution to monetary theory was to place the overall demand for money in the context of individual choices as to the fraction of one's wealth to keep on hand as ready cash. This approach, set out clearly in a manuscript of the early 1870s, was developed by Marshall's Cambridge successors, especially A.C. Pigou and F. Lavington, into what is termed the "Cambridge k" approach. It laid the background for the treatment of the demand for money in J.M. Keynes. On international monetary theory, Marshall espoused a form of purchasing power parity.

Marshall's name is particularly associated with his proposals for "symmetalism," the use of a fixed-weight combination of gold and silver as the monetary base, and for indexed contracts based on a "tabular standard of value," or price index, to be maintained by the government. The former was offered as an improvement on fixed-ratio bimetalism, of which he was never more than a lukewarm adherent. Marshall had interesting, if fragmentary, insights into business fluctuations and general unemployment, which he viewed as temporary disequilibrium consequences of credit market dislocations. These spilled over into general coordination failures, with unemployment in one market spreading to others by reducing demand in cumulative fashion–the germ at least of the multiplier concept. On the other hand, Say's law was maintained as an equilibrium truth of great importance. He saw the remedies for cyclical unemployment in the "continuous adjustment of means to ends, in such a way that credit can be based on the solid foundation of fairly accurate forecasts," and in curbs on reckless inflations of credit which are "the chief cause of all economic malaise".

9.1.1. International Trade

Marshall's major contribution to international trade theory was his well-known geometrical analysis of the equilibrium and stability of two-country trade by means of intersecting offer curves. Each country's offer curve indicated the number of "bales" of home goods it was prepared to exchange for a specified number of bales of foreign goods, demand being elastic or inelastic as an increase in the latter caused the former to increase or decrease. Possibilities of multiple and unstable offer-curve intersections were noted. The offer curves themselves were taken as data, although complex readjustments of production and consumption underlay them. The need for a separate theory of international trade was justified, in classical vein, by the supposed international immobility of factors which remain mobile domestically. The main purpose of this theoretical apparatus was to examine the effects of tariffs. A country might gain by selfishly exploiting its monopoly power through restricting trade, and would certainly gain if trading equilibrium was on an inelastic portion of the foreign offer curve. But Marshall came increasingly to doubt the transferability of this result to a multi-country case, although admitting that it might apply to an export tax on an exceptional commodity (like British steam coal) lacking close substitutes and incapable of being produced elsewhere. A related attempt to construct a theoretical measure of the "net benefit" a country gains from foreign trade, analogous to the measures of consumer and producer surplus, was not entirely satisfactory as the partial equilibrium context had clearly been transcended. On matters of concrete trade policy for Britain, Marshall was a firm but cautious adherent of free trade, even unilateral free trade, but became increasingly concerned with the prospects for Britain's position in the world economy. The discussion in *Industry and Trade* of the links between foreign competition and domestic industrial organization and structure reflected this concern.

9.1.2. Concept of National Income

It was Marshall who developed the concept of the national income, which he defined as follows. "The labour and capital of the country acting on its natural resources, produces annually a certain net aggregate of commodities-material and immaterial, including services of all kinds". This net output is called the national income. He believed that national income depends upon the co-operation of all the factors of production. Marshall distinguished between land, labour and capital. He held that capital is the result of man working upon natural resources. Hence there are only two essential agents (factors) of production- man (labour) and nature (land). While discussing the role played by the different agents of in production, Marshall pointed out that these factors were both competitive and complementary. Though they competed for employment during the process of production in the short run, in the long run all these factors would complement one another.

In the field of distribution Marshall made use of the forces of supply and demand. He held that each factor would be used in production to the extent where its marginal productivity was equal to its marginal cost. The marginal productivity determined the demand price of the factor and the marginal cost determined its supply price. However, this is not applicable in the determination of land rent. Marshall divided cost into prime costs and supplementary cost. Prime cost, which is now called variable cost, includes things like labour wages and raw materials. This cost changes in the short run along with production. A supplementary cost, which is now called fixed, cost, include things like depreciation of the plant, interest on loans, rent, salaries of top executives etc. This does not change in the short run.

In his analysis of marginal cost in relation to value, Marshall restated the principle of substitution. He stated that in the modern world production has become very complicated and nearly all the factors of production is controlled by businessmen and employers who had acquired specialized knowledge in organising different economic processes. Then, the employer will choose those factors which are profitable to him. In other words, the employer will always substitute a costly factor by a less costly factor. This principle is also applicable to methods of production. If there are two methods of producing a commodity -by skilled and unskilled, then efficient in production to its cost. Marshall has made some original contributions to monetary economics. Marshall held that value of money is a function of its supply and demand for it. He put forward the 'purchasing power-parity' theory to explain the rate of exchange between countries with mutually inconvertible currencies. Besides, Marshall introduced the chain method of index numbers, proposed paper currency based on gold and silver symmetallism and distinguished between real rate of interest and money rate of interest.

9.2. The Aggregative Strand of Marshall's Thought

Though Marshall's attention was primarily directed to micro-economic problems, aggregative themes still occupied a place in his thought. In his view, the major micro-economic question was the determination of the general price level. Short term fluctuations in output and employment were peripheral matters; when they occurred they were expected to be temporary and slight. His analysis of the general price level was developed around a version of the 'quantity theory' of money. Much of the earlier discussion of this point of doctrine had proceeded from the tautological statement that the quantity of money multiplied by the number of times it was spent in a given time period (the velocity of circulation) would necessarily be equal to the average price level multiplied by the total number of transactions; this expression, after all, amounted to no more than two ways of viewing total expenditure. Marshall modified this procedure by shifting the focus from the rate at which the money supply turned over to an examination of the money balances held by the community. In the hands of one of Marshall's pupils this way of viewing money was later to open up fresh analytical horizons. Marshall's own results from the use of this 'cash balance' approach, however, were essentially no different from those that had been reached via the 'velocity of circulation' route. He maintained that the amount of money held was regulated by the institutional arrangements of the economy and, on ceteris paribus assumptions could be treated as a constant.

The effect of this procedure was to reinforce the essential requirement of Say's Law: that all income would be spent. The possibility of leakages into idle balances could, for practical purposes, be ignored. Money was interesting primarily in relation to spending and to the general price level, rather than for any connexion it might bear to the level of interest rates. This conclusion, of course, gained additional strength from Marshall's insistence - which was common to the neoclassical tradition as a whole - that interest rates would be established through the interaction of the supply of loanable funds (fed by saving) and the demand for loanable funds (stimulated by the productivity of capital). Moreover, the rate of interest could be relied upon to produce an equilibrium between decisions to save and to invest. Should the demand for loanable funds increase, the rate of interest would rise, making it more attractive for people to reduce consumption spending and to save. Conversely; should the public choose to save more, the rate of interest would fall. Investors would then be induced to increase both their borrowings and their expenditures on plant and equipment. Further, this way of looking at the matter implied that the intersection of the curves of supply and demand for loanable funds determined the equilibrium rate of interest. The position of these curves, in turn, was established by the thriftiness of the community (on the supply side) and by the productivity of capital (on the demand side). This line of reasoning, while supporting Say's Law interpretation of aggregative economic activity, did not preclude the possibility of economic instability. Though no disturbances of the scale experienced in the 1930s clouded the horizons of Marshall and his neoclassical contemporaries, they did observe modest cycles of boom and bust. How were these fluctuations to be explained? In Marshall's view the main answer was to be found in the psychology of the business community. Waves of optimism and pessimism seemed endemic to it. When business men were bullish the demand for loans increased. This phase might generate capital spending on many high risk undertakings, some of which were doomed to failure. And when they did fail the bubble was pricked. Pessimism replaced optimism as the dominant mood; investment and economic activity generally would be curtailed.

As Marshall described the process: The recent history of fluctuations of general credit shows much variety of detail, but a close uniformity of general outline. In the ascending phase, credit has been given somewhat boldly, and even to men whose business capacity has not been proved. For, at such times a man may gain a profit on nearly every transaction, even though he has brought no special knowledge or ability to bear on it; and his success may probably tempt others of like capacity with himself, to buy speculatively. If he is quick to get out of his ventures, he probably makes a profit. But his sales hasten a fall of prices, which must have come in the course of time. Though the fall is likely to be slight at first; yet each downward movement impairs the confidence which had caused the rise of prices, and is still giving them some support. The fall of a lighted match on something that smoulders has often started a disastrous panic in a crowded theatre. Credit cycles, however, were still incapable of converting 'partial' over-production into 'general' over production. Given time, the economic system would adjust itself to its normal full employment level of operations. No special action on the part of government was required to accomplish this result and, indeed, direct government intervention might make matters worse. The tendency toward instability could, however, be moderated by anticipatory action on the part of the monetary authorities whose proper role was to minimize discrepancies between prevailing interest rates and the rate that would be established through the normal interplay of the supply of and demand for loanable funds. The main thrust of Marshall's aggregative analysis thus buttressed faith in the capability of the economic system, if left alone, to eradicate involuntary idleness. In the final edition (1920) of his Principles, however, Marshall added one dark hint that the analytical basis for his conclusion might ultimately need to be revised. Following an orthodox neoclassical discussion of the relationship between productivity, thrift, and the rate interest, he inserted a note of qualification, everyone understands generally the causes which have kept the supply of accumulated wealth so small relatively to the demand for its use, that that use is on balance a source of gain, and can therefore require payment when loaned. Everyone is aware that the accumulation of wealth is held in check, and the rate of interest so far sustained, by the preference which the great mass of humanity have for present over deferred gratifications, or, in other words, by their unwillingness to wait. And indeed the true work of economic analysis in this respect is, not to emphasize this familiar truth, but to point out how much more numerous are the exceptions to this general preference than would appear at first sight.

9.3. Marshall and Economic Policy

By his own account, Marshall was originally attracted to the serious study of economics by a desire to understand the causes of poverty and the means by which it could be alleviated. He emerged from his investigation convinced that: the social and economic forces already at work are changing the distribution of wealth for the better: that they are persistent and increasing in strength; and that their influence is for the greater part cumulative; that the socio-economic organism is more delicate and complex than at first sight appears; and that large ill-considered changes might result in grave disaster.His sympathy for the sufferings of the mass of humanity had by no means diminished. But these impulses were now substantially tempered by the belief that radical measures to alter the existing economic order would be unwise. In particular, he opposed a socialistic programme on the grounds that: the collective ownership of the means of production would deaden the energies of mankind, and arrest economic progress; unless before its introduction the whole people had acquired a power of unselfish devotion to the public good which is now relatively rare. And, it might probably destroy much that is most beautiful and joyful in the private and domestic relations of life. These are the main, reasons which cause patient students of economics generally to anticipate little good and much evil from schemes for sudden and violent reorganization of the economic, social and political conditions of life. While the market system as portrayed by Marshall was largely benevolent, his analysis also demonstrated that in certain situations unregulated markets could not be relied upon to yield socially desirable results. Prominent among the exceptions were the cases in which - for technical reasons - competition would prove to be wasteful and inefficient, if not a practical impossibility. The 'natural monopolies' (a term which Marshall associated primarily with such public services as water supply, power generation, etc.) could not usefully be organized in accordance with the competitive plan and the case for government regulation (if not public ownership) in these instances was clear. He was reluctant, however, to recommend government intervention in those sectors of the economy in which increasing returns to scale threatened to produce industrial concentrations, even though these circumstances implied that individual firms could enjoy considerable market power and that prices would not be competitively determined. This problem, he contended, deserved continuing study. His general position on the life cycle of firms led him to the conclusion that the potential market power of large business units was unlikely to be abused for long.

Though he was disposed to view the market as a sensitive instrument through which an economy's resources could be efficiently allocated, he also recognized that its performance could be improved. For this purpose, improvements in public education were particularly important. Consumers and producers could then conduct their affairs more intelligently by enhancing the rationality of their choices. Moreover, improved economic education would do much to eradicate one of the blights of an unregulated market system - the bouts of speculation which gave rise to harmful fluctuations.

Marshall was also prepared to entertain the possibility that governments could play a useful role in improving the allocative efficiency of markets. Would not the sum of social satisfactions be increased, he asked, if the productive resources of society were shifted in favour of lines of production subject to increasing returns and away from those in which decreasing returns prevailed? Greater outputs could then be obtained from the existing stock of resources. Governments could encourage a re-allocation of resources along these lines through appropriate taxes and subsidies. He advanced this suggestion, however, with the utmost caution, pointing out that such policies could be justified only when it could be shown that the gains in satisfactions arising from expanded outputs in the subsidized sectors more than offset the losses in utility associated with higher tax levies on others. He recognized that this criterion would be difficult to apply to practical cases. Conceivably the introduction of maximization of aggregate utility as a goal of public policy could also be used to support recommendations for a redistribution of income. If it could be assumed that the marginal utility of money was likely to be greater far a poor man than for a rich one, it followed that society's aggregate satisfactions would be enlarged through a redistribution from rich to poor. Marshall did not draw this conclusion. He did recommend a less systematic scheme of income redistribution for further study when he wrote of the possibilities of 'economic chivalry'. Such a regime would tax the rich to ameliorate the distress of those still trapped in poverty.

When Marshall passed away in 1924, Keynes described him 'the greatest economist in the world for a hundred years'. His influence on the development of modern economics has been so great that he is considered as the father of modern economics. His remarkable work 'Principles of Economics' guided the economic literature of more than a generation. The works of Pigou, Edgeworth, Keynes, Mourice Dobb and many other economists is a proof for this. However, there are some limitations in Marshall's analysis. Certain logical inconsistencies can be found in Marshall's theories. As a result he could not complete his synthesis of classical and marginal economics. For instance, his synthesis of marginal utility and cost of production are considered to be incomplete by some critics. The most important criticisms are:

- ✰ Many of his concepts like consumer surplus, representative firm etc are vague.
- ✰ He almost completely neglected macro-economics.
- ✰ He gave undue importance to perfect competition and omitted the treatment of large scale corporations; cartels etc. Some critics point out these limitations and argues that the Marshallian system has crumbled down because since the publication of his work, advances have been made in economic theory. But they forget the fact that this advancement has been made possible by the solid foundations laid by Marshall himself.

10
Neoclassical Economics

10.1. Neoclassical Economics

Economists publicly disagree with each other so often that they are easy targets for standup comedians. Yet non-economists may not realize that the disagreements are mostly over the details –the way in which the big picture is to be focused on the small screen. when it comes to broad economic theory, most economists agree. President Richard Nixon, defending deficit spending against the conservative charge that it was "Keynesian," is reported to have replied, "we're all keynesians now." In fact, what he should have said is "we're all neoclassicals now, even the keynesians," because what is taught to students, what is mainstream economics today, is neoclassical economics. By the middle of the nineteenth century, english-speaking economists generally shared a perspective on value theory and distribution theory. the value of a bushel of corn, for example, was thought to depend on the costs involved in producing that bushel. the output or product of an economy was thought to be divided or distributed among the different social groups in accord with the costs borne by those groups in producing the output. this, roughly, was the "classical theory" developed by Adam Smith, David Ricardo, Thomas Robert Malthus, John Stuart Mill, and Karl Marx. But there were difficulties in this approach. chief among them was that prices in the market did not necessarily reflect the "value" so defined, for people were often willing to pay more than an object was "worth." the classical "substance" theories of value, which took value to be a property inherent in an object, gradually gave way to a perspective in which value was associated with the relationship between the object and the person obtaining the object. Several economists in different places at about the same time (the 1870s and 1880s) began to base value on the relationship between costs of production and "subjective elements," later called "supply" and "demand." This came to be known as the marginal revolution in economics, and the overarching theory that developed from these ideas came to be called neoclassical economics. (The first to use the term "neoclassical economics" seems to have been the american economist thorstein veblen). The framework of neoclassical economics is easily summarized. Buyers attempt to maximize their gains from getting goods, and they do this by increasing their purchases of a good until what they gain from an extra unit is just balanced by what they have to give up to obtain it. In this way they maximize "utility"-the satisfaction associated with the consumption of goods and services. Likewise, individuals provide labour to firms that wish to employ them, by balancing the gains from offering the marginal unit of their services (the wage they would receive) with the disutility of labour itself-the loss of leisure. Individuals make choices at the margin. This results in a theory of demand for goods, and supply of productive factors. Similarly, producers attempt to produce units of a good so that the cost of producing the incremental or marginal unit is just balanced by the revenue it generates. in this way they maximize profits. Firms also hire employees up to the point that the cost of the additional hire is just balanced by the value of output that the additional employee would produce. The neoclassical vision thus involves economic "agents," be they households or firms, optimizing (doing as well as they can), subject to all relevant constraints. Value is linked to unlimited desires and wants colliding with constraints, or scarcity. the tensions, the decision problems, are worked out in markets. Prices are the signals that tell households and firms whether their conflicting desires can be reconciled. At some price of cars, for example, I want to buy a new car. At that same price others may also want to buy cars. but manufacturers may not want to produce as many cars as we all want. Our frustration may lead us to "bid up" the price of cars, eliminating some potential buyers and encouraging some marginal producers. As the price changes, the imbalance between buy orders and sell orders is reduced. This is how optimization under constraint and market interdependence leads to an economic equilibrium. This is the neoclassical vision. neoclassical economics is what is called a meta-theory. That is, it is a set of implicit rules or understandings for constructing satisfactory economic theories. It is a scientific research program that generates economic theories. its fundamental assumptions are not open to discussion in that they define the shared understandings of those who call themselves neoclassical economists, or economists without any adjective. those fundamental assumptions include the following:

- ✰ People have rational preferences among outcomes.
- ✰ Individuals maximize utility and firms maximize profits.
- ✰ People act independently on the basis of full and relevant information. Theories based on, or guided by, these assumptions are neoclassical theories. Thus, we can speak of a neoclassical theory of profits, or employment, or growth, or money. We can create neoclassical production relationships between inputs and outputs, or neoclassical theories of marriage and divorce and the spacing of births. Consider layoffs, for example. a theory which assumes that a firm's layoff decisions are based on a balance between the benefits of laying off an additional worker and

the costs associated with that action will be a neoclassical theory. a theory that explains the layoff decision by the changing tastes of managers for employees with particular characteristics will not be a neoclassical theory. What can be contrasted to neoclassical economics? some have argued that there are several schools of thought in present-day economics. they identify (neo-)Marxian economics, (neo-)Austrian economics, post-keynesian economics, or (neo-)institutional economics as alternative matheoretical frameworks for constructing economic theories. To be sure, societies and journals promulgate the ideas associated with these perspectives. some of these schools have had insights that neoclassical economists have learned from; the austrian insights on entrepreneurship are one example. But to the extent these schools reject the core building blocks of neoclassical economics–as austrians reject optimization, for example–they are regarded by mainstream neoclassical economists as defenders of lost causes or as kooks, misguided critics, and antiscientific oddballs. The status of non-neoclassical economists in the economics departments in english-speaking universities is similar to that of flatearthers in geography departments: it is safer to voice such opinions after one has tenure, if at all. one specific attempt to discredit neoclassical economics developed from British Economist Joan Robinson and her colleagues and students at cambridge in the late fifties and early sixties. The so-called two Cambridges capital controversy was ostensibly about the implications, and limitations, of Paul Samuelson and Robert Solow's aggregating "capital" and treating the aggregate as an input in a production function. However, this controversy really was rooted in a clash of visions about what would constitute an "acceptable" theory of the distribution of income. what became the post-keynesian position was that the distribution of income was "best" explained by power differences among workers and capitalists, while the neoclassical explanation was developed from a market theory of factor prices. eventually the controversy was not so much settled as laid aside, as neoclassical economics became mainstream economics. How did such orthodoxy come to prevail? In brief, the success of neoclassical economics is connected to the "scientificization" or "mathematization" of economics in the twentieth century. it is important to recognize that a number of the early marginalists, economists like William Stanley Jevons and F.Y. Edgeworth in England, Leon Walras in Lausanne, and Irving Fisher in the United States, wanted to legitimize economics among the scholarly disciplines. The times were optimistic about a future linked to the successes of technology. Progress would be assured in a society that used the best scientific knowledge. Social goals would be attainable if scientific principles could organize social agendas. Scientific socialism and scientific management were phrases that flowed easily from the pens of social theorists.

Neoclassical economics conceptualized the agents, households and firms, as rational actors. Agents were modelled as optimizers who were led to "better" outcomes. the resulting equilibrium was "best" in the sense that any other allocation of goods and services would leave someone worse off. thus, the social system in the neoclassical vision was free of unresolvable conflict. The very term "social system" is a measure of the success of neoclassical economics, for the idea of a system, with its interacting components, its variables and parameters and constraints, is the language of midnineteenth- century physics. this field of rational mechanics was the model for the neoclassical framework. agents were like atoms; utility was like energy; utility maximization was like the minimization of potential energy, and so forth. in this way was the rhetoric of successful science linked to the neoclassical theory, and in this way economics became linked to science itself. whether this linkage was planned by the early marginalists, or rather was a feature of the public success of science itself is less important than the implications of that linkage. For once neoclassical economics was associated with scientific economics, to challenge the neoclassical approach was to seem to challenge science and progress and modernity.

Neoclassical economics should be of interest to political theorists for two rather different reasons. first, many issues of interest to political theorists have an economic dimension. for example, environmental measures may have costs in terms of the general well-being of individuals pursuing distributive justice will often serve to reduce that which is to be distributed; otherwise attractive policy measures may have undesirable incentive effects; and so on. Second, neoclassical economics offers both a framework for analysis of how institutions work and a substantive body of relevant propositions and empirical evidence. the domain of such institutional analysis includes, specifically, the political institutions that, at least on some readings, are the core of political philosophy's concerns.

Institutionalism is a twentieth century phenomenon that took birth in America. the social factors which lead to the rise of institutionalism may be briefly stated as follows. In the nineteenth century though American capitalism achieved tremendous success, the conditions of the working class was rather bad. Their standard of living and working conditions were miserable. to worsen the situation, towards the end of 19th century, monopolies became powerful intuitions in America and economic and political power was concentrated in the hand of big business. In the beginning there was not much reaction from the American economic thinkers against this social situation (this was partly because till the early 20th century, America had not many economic thinkers as compared to england and germany. The study of economics as a science started in America only in 1798 and the development of American economic thought was gradual) however, in early 20th century, some American economists reacted against the existing economic rules and suggested to reform the society by having social control over institutions. these groups of economists lead by thorstein veblen and his followers like commons and mitchell are referred to as institutionalists.

10.2. Important Ideas of Institutionalism

Institutional or social economics signifies the role of institutional factors like habits, customs, tastes, culture etc in economic life. the institutionalists held that economic problems must be studied with broad social and political framework. following are the basic ideas of institutionalism.

- ☆ They emphasized the role of institutions (by which we mean factors like customs, social habits laws, ways of living, modes of thinking etc.) in economic life.
- ☆ They believed that the economy must be studied as a whole, in other words, they believed that the group social behaviour is more important in analysing economic problems than the individual behavior (as emphasized by the marginalist school)
- ☆ They advocated the evolutionary approach in economic analysis. They held that the study of the evolution and functioning of economic institutions should be the central theme in economics.
- ☆ Since human behaviour is constantly changing, economic laws should always be relative to time and place.
- ☆ They rejected the idea of formal equilibrium. According to them, maladjustments in economic life under existing institutions are not departures from normal equilibrium.
- ☆ They held that important motives governing individual action cannot be measured.
- ☆ In their methodology, the institutionalists held the following views (i) they advocated inductive method rather than deductive method. in this regard, they rejected the abstract reasoning of the marginalists as unrealistic. (ii) they questioned the pleasure-pain psychology of the marginalists. (iii) they considered it necessary for the social scientist to appraise the quality of life. (iv) they differed from pure economics in their willingness to project past and present trends of social evolution into the future and to analyse the implications of such trends. institutionalism not only had its impact on economic thought, but had great influence on economic policy formulation of the modern world, especially the USA, for example, the 'new deal' policy of president roosevelt and policies regarding labour legislation, social security measures, public utility system etc were influenced by institutionalism.

10.2.1. John Maynard Keynes and Post Keynesians (1883-1946)

John Maynard Keynes was an English economist and is the author of "the general theory of employment, interest and money" (1936). as a result of the influence of this work, he became the most influential economist of the twentieth century. written during the height of the depression, it offered a new explanation of the depression and the unemployment that plagued it. Keynes' new theory probably also appealed to economists because it provided an alternative to the traditionally held view that unemployment can and should be eliminated by a drop in wage rates. Keynes new theory conveyed a politically much more pleasant solution to the problem of unemployment. According to Keynes, the solution to unemployment was a growth in government spending. The particular form of government spending advocated by keynes was for the government to purposely adopt a policy of budget deficits which he called "fiscal policy." To arrive at this seemingly simple conclusion, keynes developed a highly complex argumentation brimming with new economic terms and concepts of his own devising, such as "multipliers," "consumption and saving functions," "the marginal efficiency of capital," "liquidity preference," "I-S curve," and many others. the essence of keynes' theory, however, involves a shift from classical economics' concern with the production of wealth to a concern with the consumption of wealth. According to Keynes, say's law is not true; that is, supply does not create its own demand. Rather, according to Keynes, supply is capable of outstripping demand, with the result that goods remain unsold, and production and employment are correspondingly cut back. as a result, the solution to unemployment, according to keynes, is not to reduce wages and prices, as the classicals advocated, but to increase consumption through the spending of money by the government.

John Maynard Keynes was born in 1883 the year which marked the death of an intellectual giant, Marx. Keynes has been one of the greatest and the most controversial economists of the 20th century.he was the son of john neville keynes, a noted economist. he studied economics at cambridge under marshall and pigou. After his graduation in 1905, keynes sat for the civil service examination. it may be interesting to note that the architect of "new economics" got his worst marks in economics in his civil service examination. his comment was "the examiners presumably knew less than I did." in 1905, Keynes joined India office and served for two years. After that he went to cambridge university to teach economics. In 1912, he became editor of the economic journal in which capacity he continued until 1945. he was a member of the royal commission on indian currency and finance (1913 – 1914). He was principal representative of the treasury at paris peace conference after world war I. However, he resigned from treasury and returned to teaching. between 1909-and 1915 he was a lecturer at Kings's College, Cambridge. He served the British Treasury from 1915 to 1919. he also served as a Director of the Bank of England. He visited the united states in 1934 and was influential in bringing about the new deal recovery programme of president roosevelt. he was the leader of the British delegation to the Bretten Woods Conference (1944) and served as a Governor of the I.B.R.D. and I.M.F. his book, "The General Theory of Employment, Interest and Money" marks a turning point in the history of economic thought. Hence, the term "keynesian revolution" is often applied to describe the economic ideas of Keynes. Keynes married Lydia Lopokova, a russian in 1925. He died in 1946 at the age of 62. The important works of Keynes are Indian currency and finance (1913), the economics of war in Germany (1915), the economic consequences of the peace (1919), a treatise on probability(1921), a tract on monetary reform (1923), the end of laissez-faire (1926), a treatise on money (1930), the end of the gold standard (1931), the general theory of employment, interest and money (1936), how to pay for the war (1940).

10.2.2. Keynesian Economics

With the recent economic crisis, there has been much talk of John Maynard Keynes and his Economics. Keynes figured out the causes of the great depression and in doing so revolutionized thenfield of economics. According to Keynes, classical economics (basically that summarized by Alfred marshall, including Ricardo, Mill, Edgeworth, and Pigou) addresses only a special case of the economy, while his is a more general theory. the classical theory of employment says the labour market is just another market: people get paid what they make and people only work if they get paid enough to make it worth it. Since it seemed unlikely that society had run out of money-making jobs, it was assumed that unemployment was caused either by people not knowing where the jobs were (frictional unemployment) or insisting on being paid more than they could make (voluntary unemployment). The classicals believed in say's law which state that supply creates its own demand. If there are people around willing to work, jobs will spring up to make use of them. If people are unemployed, it must be because they're refusing to take the job. We may explain some of the tenets of keynesian economics. Keynesian economics is the view that in the short run, output is influenced by aggregate demand (total spending in the economy). In the keynesian view, aggregate demand does not necessarily equal the productive capacity of the economy. Keynes contrasted his approach to the 'classical' economics that preceded his book. Keynesian economists argue that private sector decisions sometimes lead to inefficient macroeconomic outcomes. so it requires active policy (fiscal policy) by the government, in order to stabilize output over the business cycle. Keynesian economics advocates a mixed economy – predominantly private sector, but with a role for government intervention during recessions. The advent of the global financial crisis in 2008 has caused resurgence in keynesian thought. Prior to the publication of Keynes's general theory, mainstream economic thought was that the economy existed in a state of general equilibrium. This perception is reflected in say's law. Keynes's Theory was significant because it reversed the mainstream thought of the time and brought about a greater awareness that problems such as unemployment are not a product of laziness, but the result of a structural inadequacy in the economic system. He argued that there was no guarantee that the goods that individuals produce would be met with demand and unemployment was a natural consequence. He saw the economy as unable to maintain itself at full employment and believed that it was necessary for the government to step in and put under-utilized savings to work through government spending. Keynesians, therefore advocate an active stabilization policy to reduce the amplitude of the business cycle. According to the theory, government spending can be used to increase aggregate demand, thus increasing economic activity, reducing unemployment and deflation.

10.2.3. Keynesian Theory of Employment

The most important theory of Keynes was the theory of employment. keynes criticized and rejected the classical theory of employment which denied the existence of unemployment or over production in the economy. The starting point of Keynesian theory of employment is the concept of effective demand. According to keynes employment depends on effective demand and hence unemployment is due to lack of effective demand. Aggregate demand, total income, total output and employment according to Keynes, employment depends on aggregate demand which is equal to aggregate income. As such we can say that the general theory of employment is the theory of aggregate income or output. Employment results in the production of output or income. in other words, total output or employment is determined by aggregate demand and aggregate supply. Aggregate demand consists of consumption demand plus investment demand plus government demand (expenditure). Consumption function and investment multiplier.

Keynes developed the concept of multiplier which was first developed by Richard F. Kahn in 1931. It is the multiple increases in income as a result of an increase in investment. Exogenous increases in government spending increases total spending by a multiple of that increase. The consumption function is a single mathematical function used to express consumer spending. It was developed by John Maynard Keynes and detailed most famously in his book *the general theory of employment, interest, and money*. The function is used to calculate the amount of total consumption in an economy. It is made up of autonomous consumption that is not influenced by current income and induced consumption that is influenced by the economy's income level. This function can be written in a variety of ways, an example being. This is probably the most simplistic form of the consumption function. The simple consumption function is shown as the affine function: where

c = total consumption,

c0 = autonomous consumption

c1 is the marginal propensity to consume (the induced consumption), and

yd = disposable income (income after government intervention – benefits, taxes and transfer

Payments – or y + (g – t)). Autonomous consumption represents consumption when income is zero. In estimation, this is usually assumed to be positive. The marginal propensity to consume (mpc), on the other hand measures the rate at which consumption is changing when income is changing. In a geometric fashion, the mpc is actually the slope of the consumption function. The mpc is assumed to be positive. Thus, as income increases, consumption increases. However, according to him, consumption increases with income, but the increase in consumption is less than the increase in income. The Keynesian consumption function is also known as the absolute income hypothesis, as it only bases consumption on current income and ignores potential future income.

Investment expenditure refers to the creation of new assets *i.e.* an addition to the stock of existing capital assets. According to Keynes investment demand depends upon two factors – the expected rate of profit and the rate of interest. The expected rate of profit is known as the marginal efficiency of capital (mec). Investment demand increases with the increase in the expected rate of profit and decreases with the increase in the rate of interest. Symbolically, Keynes investment demand is expressed as:

id = f (mec, ir) **effective demand = total output = total income = employment marginal efficiency of capital** the marginal efficiency of capital (mec) is that rate of discount which would equate the price of a fixed capital asset with its present discounted value of expected income. The term "marginal efficiency of capital" was introduced by john Maynard Keynes in his general theory, and defined as "the rate of discount which would make the present value of the series of annuities given by the returns expected from the capital asset during its life just equal its supply price". In other words, it is the annual percentage return on the additional unit of capital. It represents the market rate of interest.

10.3. Rate of Interest

According to Keynes, interest is the reward for parting with liquidity. Liquidity preference is the preference of the people to keep their assets in the form of money. He argued that saving and investments are not the main determinants of interest rates, especially in the short run. According to him, the rate of interest is determined by the supply of money (Central Bank of a Country) and the demand for money (liquidity preference due to transaction, precautionary and speculative motives). Keynes developed the liquidity preference theory of interest according to which liquidity preference and supply of money determines the rate of interest.

Classical economists have traditionally yearned for balanced government budgets. Keynesians, on the other hand, believe this would aggravate the underlying problem. Keynes's view was that insufficient buying-power caused the Depression. Keynes developed a theory which suggested that active government policy could be effective in managing the economy. Keynes advocated counter cyclical fiscal policies, that is, policies that acted against the tide of the business cycle. That is deficit spending when a nation's economy suffers from recession or unemployment is persistently high–and the suppression of inflation in boom times by either increasing taxes or cutting back on government outlays. He argued that governments should solve problems in the short run rather than waiting for market forces to do it in the long run, because, "in the long run, we are all dead."

10.3.1. Keynes and Classicals: A Comparison

Classical and Keynesian economics are two important schools of thought. Each school takes a different approach to the economic study of monetary policy, consumer behavior and government spending. A few basic distinctions separate these two schools. Classical economic theory is rooted in the concept of a laissez-faire economic market. A Laissez-Faire– also known as free–market requires little to no government intervention. It also allows individuals to act according to their own self interest regarding economic decisions. This ensures economic resources are allocated according to the desires of individuals and businesses in the marketplace. Keynesian economic theory relies on spending and aggregate demand. Keynesian economists believe the aggregate demand is influenced by public and private decisions. Public decisions represent government agencies and Private decisions include individuals and. Government spending is not a major force in a classical economic theory. Classical economists believe that consumer spending and business investment represents the more important parts of a nation and economic growth. Too much government spending takes away valuable economic resources needed by individuals and businesses. To classical economists, government spending and involvement can retard a nation's economic growth. Keynesian economics relies on government spending to a nation's economic growth during sluggish economic downturns. Similar to classical economists, Keynesians believe the nation's economy is made up of consumer spending, business investment and government spending. However, Keynesian theory dictates that government spending can improve economic growth in the absence of consumer spending or business investment. Classical economics focuses on creating long-term solutions for economic problems. The effects of inflation, government regulation and taxes can all play an important part in developing classical economic theories. Classical economists also take into account the effects of other current policies and how new economic theory will improve or distort the free market environment. Keynesian economics often focuses on immediate results in economic theories. Policies focus on the short-term needs and how economic policies can make instant corrections to a nation's economy. This is why government spending is such a key component of Keynesian economics. During economic recessions and depressions, individuals and businesses do not usually have the resources for creating immediate results through consumer spending or business investment. The government is seen as the only force to end these downturns through fiscal policies providing instant economic results. Classical follow the basic assumption that economy is in full employment. The wages and prices are very flexible and there is no need of fiscal or monetary policy. The invisible hand makes the economy self correctable. The Aggregate supply curve is vertical according to Classical and so any rise in aggregate demand will increase prices and not production. In Classical Theory, Government has minimal role in the economy, and the macro-economy is self adjusting; meaning consumers and businesses will correct any problems with the economy automatically over time. Classical theory focuses on long-term goals. Keynesian economics follow the basic assumptions that Economy may not be in full employment in the short run. Wages are rigid and prices are sticky. Fiscal policy may be needed to correct the disequilibrium or improve the efficiency of the economy. Aggregate supply is upward sloping in the short run so a rise in aggregate demand may raise the production. In Keynesian theory, Government has a large role in the economy, and focuses on short-term goals.

10.3.2. Keynes as a Critic of Classical Economics

Classical economists generally include Adam smith, Jeremy Bentham, T.R.Malthus, David Ricardo, J.S.Mill, J.B.Say etc. But Keynes included, the neoclassical economists including Marshall, Pigou and Edgeworth in the category of classical economists. Hence, Keynes criticised the economic theories developed by these writes mainly Pigou. The main points of contrast between the Keynesian and Classical theories of Income and Employment are discussed in brief as under.

1. Unemployment

The classical economists explained unemployment using traditional partial equilibrium supply and demand analysis. According to them unemployment results when there is an excess supply of labour at a particular higher wage level. By accepting lower wage, the unemployed workers will go back to their jobs and the equilibrium between demand for labour and supply of labour will be established in the labour market in the long period. This equilibrium in the economy is always associated with full employment level. According to classical economists unemployment results when the wage level of workers is above the equilibrium wage level and as a result there of, the quantity of labour supplied is higher than quantity of labour demanded. The difference between the two (supply and demand) is unemployment. J. M Keynes and his followers, however reject the fundamental classical theory of full employment equilibrium in the economy. They consider it as unrealistic. According to them full employment is a rare phenomenon in the capitalistic economy. The unemployment occurs they say when the aggregate demand function intersects the aggregate supply function at a point of less than full employment level. Keynes suggested that in the short period, the government can raise aggregate demand in the economy through public investment programmes to reduce unemployment.

2. Say's Law of Market

Say's Law "Supply creates its own demand" is central to the classic vision of the economy.

According to say the production of goods and services generates expenditure sufficient to ensure that they are sold in the market. There is no deficiency of demand for goods and hence no need t unemployed workers. According to him full employment is a normal condition of market economy. J. M Keynes has strongly refuted. Say's Law of Market with the help of effective demand. Effective demand is the level of aggregate demand which is equal to aggregate supply. Whenever there is efficiency in aggregate demand a part of the goods produced remain unsold in the market which lead to general over production of goods and services in the market. When all the goods produced in the market are not sold, the firms lay off workers. The deficiency in demand for goods creates unemployment in the economy.

3. Equality between Saving and Investment

Classical economists are of the view that saving and investment are equal at the full employment level. If at any time flow of savings is greater than flow of investment, then the rate of interest declines in the money market this leads to an increase in investment. The process continues till the flow of investment equals the flow of saving. Thus according to the classical economists, the quality between saving and investment is brought about through mechanism of rate of interest. J.M.Keynes is however of the view that equality between saving and investment is brought about through changes income rather than the changes in interest rate.

4. Money and Prices

The classical economists are of the opinion that price level varies in response to changes in the quantity of money. The quantity theory of money seeks to explain the value of money in terms of changes in its quantity. J.M.Keynes has rejected the simple quantity theory of money. According to him if there is recession in the economy and the resources are lying idle and unutilized an increased spending of money lead to substantial increase in real output and employment without affectively price level.

5. Marx and Keynes

There are some similarities and more dissimilarities between Marx and Keynes. Both of them were interested in macroeconomics. Both of them regarded inadequate demand as the cause of trade cycles. Both emphasized the falling rate of profit. Both believed that the rewards for capitalist groups were excessive.

6. Dissimilarities between Marx and Keynes

Keynes thought that capitalist system could be preserved through the modifications of the capitalist institutions and changes in policy, Marx predicted the inevitable decline and fall of capitalism. While Marx became the 'prophet of doom', Keynes became the 'prophet of boom'. Keynes was the very opposite of Marx and Keynesian philosophy should be regarded as a major alternative to Marxism. Of course, Keynes conceded that Marx had some idea of the unemployment problem in a capitalist society.

10.4. New Classical Macroeconomics

New classical economics introduced a set of macroeconomic theories that were based on optimising microeconomic behavior. These models have been developed into the Real Business Cycle Theory, which argues that business cycle fluctuations can to a large extent be accounted for by real (in contrast to nominal) shocks. Beginning in the late 1950s new classical macroeconomists began to disagree with the methodology employed by Keynes and his successors. New classical theorists demanded that macroeconomics be grounded on the same foundations as microeconomic theory, profit-maximizing firms and rational, utility-maximizing consumers.

Monetarism is a school of economic thought that emphasizes the role of monetary authority (Central Bank) in controlling the amount of money in circulation for stability and growth. It is the view within monetary economics that variation in the money supply has major influences on national output in the short run and the price level over longer periods and those objectives of monetary policy are best met by targeting the growth rate of the money. Monetarism today is mainly associated with the work of Milton Friedman, who was among the generation of economists to accept Keynesian economics and then criticize Keynes' theory of gluts based on a policy of government intervention. Thus, Monetarism is an economic theory which focuses on the macroeconomic effects of the supply of money and central banking. It argues that excessive expansion of the money supply is inherently inflationary, and that monetary authorities should focus solely on maintaining price stability.

Friedman and Anna Schwartz wrote an influential book, A Monetary History of the United States, 1867- 1960, and argued that "inflation is always and everywhere a monetary phenomenon." It attributed deflationary spirals to the reverse effect of a failure of a central bank to support the money supply during a liquidity crunch. Friedman advocated, a central bank policy aimed at keeping the supply and demand for money at equilibrium, as measured by growth in productivity and demand. Friedman focused on price stability, which is the equilibrium between supply and demand for money. The rise of monetarism accelerated from Milton Friedman's

1956 restatement of the quantity theory of money. Friedman argued that the money could be described as depending on a small number of economic variables. Thus, where the money supply expanded, people would not simply wish to hold the extra money in idle money balances; *i.e.*, if they were in equilibrium before the increase, they were already holding money balances to suit their requirements, and thus after the increase they would have money balances surplus to their requirements. These excess money balances would therefore be spent and hence aggregate demand would rise. Similarly, if the money supply were reduced people would want to replenish their holdings of money by reducing their spending. In this, Friedman challenged a simplification attributed to Keynes suggesting that "money does not matter." Thus the word 'monetarist' was coined. Milton Friedman and Anna Schwartz in their book A Monetary History of the United States, 1867-1960 argued that the Great Depression of 1930 was caused by a massive contraction of the money supply and not by the lack of investment Keynes had argued. They also maintained that post-war inflation was caused by an over-expansion of the money supply. Monetarists of the Milton Friedman school of thought believed in the 1970s and 1980s that the growth of the money supply should be based on certain formulations related to economic growth. As such, they can be regarded as advocates of a monetary policy based on a "quantity of money" target.

There was debate between Monetarists and Keynesians in the 1960s over the role of government in stabilizing the economy. Both Monetarists and Keynesians are in agreement over the fact that issues such as business cycles, unemployment, inflation are caused by inadequate demand, and need to be addressed, but they had fundamentally different perspectives on the capacity of the economy to find its own equilibrium and as a consequence the degree of government intervention that is required to create equilibrium. Keynesians emphasized the use of discretionary fiscal policy and monetary policy, while monetarists argued the primacy of monetary policy, and that it should be rules based. The debate was largely resolved in the 1980s. Since then, economists have largely agreed that central banks should bear the primary responsibility for stabilizing the economy, and that monetary policy should largely follow the Taylor rule – which many economists credit with the Great Moderation. The Global Financial Crisis, however, has convinced many economists and governments of the need for fiscal interventions and highlighted the difficulty in stimulating economies through monetary policy alone during a liquidity trap.

Rational expectations is a hypothesis in economics which states that agents' predictions of the future value of economically relevant variables are not systematically wrong in that all errors are random. Equivalently, this is to say that agents' expectations equal true statistical expected values. An alternative formulation is that rational expectations are model-consistent expectations, in that the agents inside the model assume the model's predictions are valid. The rational expectations assumption is used in many contemporary macroeconomic models, game theory and applications of rational choice theory. Since most macroeconomic models today study decisions over many periods, the expectations of workers, consumers and firms about future economic conditions are an essential part of the model. How to model these expectations has long been controversial, and it is well known that the macroeconomic predictions of the model may differ depending on the assumptions made about expectations To assume rational expectations is to assume that agents' expectations may be individually wrong, but are correct *on average*. In other words, although the future is not fully predictable, agents' expectations are assumed not to be systematically biased and use all relevant information in forming expectations of economic variables.

This way of modeling expectations was originally proposed by John F. Muth and later became influential when it was used by Robert Lucas, Jr. and others. Modeling expectations is crucial in all models which study how a large number of individuals, firms and organizations make choices under uncertainty. For example, negotiations between workers and firms will be influenced by the expected level of inflation, and the value of a share of stock is dependent on the expected future income from that stock. Rational expectations theory defines this kind of expectations as being identical to the best guess of the future (the optimal forecast) that uses all available information. Thus, it is assumed that outcomes that are being forecast do not differ systematically from the market equilibrium results. As a result, rational expectations do not differ systematically or predictably from equilibrium results. That is, it assumes that people do not make systematic errors when predicting the future, and deviations from perfect foresight are only random. In an economic model, this is typically modeled by assuming that the expected value of a variable is equal to the expected value predicted by the model. For example, suppose that P is the equilibrium price in a simple market, determined by supply and demand. The theory of rational expectations says that the actual price will only deviate from the expectation if there is an 'information shock' caused by information unforeseeable at the time expectations were formed. Rational expectations theories were developed in response to perceived flaws in theories based on adaptive expectations. Under adaptive expectations, expectations of the future value of an economic variable are based on past values. For example, people would be assumed to predict inflation by looking at inflation last year and in previous years. Under adaptive expectations, if the economy suffers from constantly rising inflation rates (perhaps due to government policies), people would be assumed to always underestimate inflation. This may be regarded as unrealistic - surely rational individuals would sooner or later realize the trend and take it into account in forming their expectations.

The hypothesis of rational expectations addresses this criticism by assuming that individuals take all available information into account in forming expectations. Though expectations may turn out incorrect, they will not deviate systematically from the expected values. The rational expectations hypothesis has been used to support some radical conclusions about economic policymaking. An example is the Proposition developed by Thomas Sargent and Neil Wallace. If the Federal Reserve attempts to lower unemployment through expansionary monetary policy economic agents will anticipate the effects of the change of policy and raise their expectations of future inflation accordingly. This in turn will counteract the expansionary effect of the increased money supply. All that the government can do is raise the inflation rate, not employment. This is a distinctly New Classical outcome. During the 1970s rational expectations appeared to have made previous macroeconomic theory largely obsolete, which culminated with the Lucas critique. However, rational expectations theory has been widely adopted throughout modern macroeconomics as a modeling assumption thanks to the work of New Keynesians such as Stanley Fischer. Rational expectations theory is the basis for the efficient market hypothesis (efficient market theory). If a security's price does not reflect all the information about it, then there exist "unexploited profit opportunities": someone can buy (or sell) the security to make a profit, thus driving the price toward equilibrium. In the strongest versions of these theories, where all profit opportunities *have been* exploited, all prices in financial markets are correct and reflect market fundamentals (such as future streams of profits and dividends). Each financial investment is as good as any other, while a security's price reflects all information about its intrinsic value.

The models of Muth and Lucas (and the strongest version of the efficient-market hypothesis) assume that at any specific time, a market or the economy has only one equilibrium (which was determined ahead of time), so that people form their expectations around this unique equilibrium. Muth's math (sketched above) assumed that P^* was unique. Lucas assumed that equilibrium corresponded to a unique "full employment" level (potential output) – corresponding to a unique NAIRU or natural rate of unemployment. If there is more than one possible equilibrium at any time then the more interesting implications of the theory of rational expectations do not apply. In fact, expectations would determine the nature of the equilibrium attained, reversing the line of causation posited by rational expectations theorists. A further problem relates to the application of the rational expectations hypothesis to aggregate behavior. It is well known that assumptions about individual behavior do not carry over to aggregate behavior. The same holds true for rationality assumptions: Even if all individuals have rational expectations, the representative household describing these behaviors may exhibit behavior that does not satisfy rationality assumptions (Janssen, 1993). Hence the rational expectations hypothesis, as applied to the representative household, is unrelated to the presence or absence of rational expectations on the micro level and lacks, in this sense, a microeconomic foundation. It can be argued that it is difficult to apply the standard efficient market hypothesis (efficient market theory) to understand the stock market bubble that ended in 2000 and collapsed thereafter; however, advocates of rational expectations say that the problem of ascertaining all the pertinent effects of the stock-market crash is a great challenge. Those studying financial markets similarly apply the efficient markets hypothesis but keep the existence of exceptions in mind.

11
New Classical Macroeconomics

11.1. New Classical Macroeconomics

New classical economics introduced a set of macroeconomic theories that were based on optimising microeconomic behavior. These models have been developed into the Real Business Cycle Theory, which argues that business cycle fluctuations can to a large extent be accounted for by real (in contrast to nominal) shocks. Beginning in the late 1950s new classical macroeconomists began to disagree with the methodology employed by Keynes and his successors. New classical theorists demanded that macroeconomics be grounded on the same foundations as microeconomic theory, profit-maximizing firms and rational, utility-maximizing consumers. Monetarism is a school of economic thought that emphasizes the role of monetary authority (Central Bank) in controlling the amount of money in circulation for stability and growth. It is the view within monetary economics that variation in the money supply has major influences on national output in the short run and the price level over longer periods and those objectives of monetary policy are best met by targeting the growth rate of the money. Monetarism today is mainly associated with the work of Milton Friedman, who was among the generation of economists to accept Keynesian economics and then criticize Keynes' theory of gluts based on a policy of government intervention. Thus, Monetarism is an economic theory which focuses on the macroeconomic effects of the supply of money and central banking. It argues that excessive expansion of the money supply is inherently inflationary, and that monetary authorities should focus solely on maintaining price stability.

Friedman and Anna Schwartz wrote an influential book, A Monetary History of the United States, 1867- 1960, and argued that "inflation is always and everywhere a monetary phenomenon." It attributed deflationary spirals to the reverse effect of a failure of a central bank to support the money supply during a liquidity crunch. Friedman advocated, a central bank policy aimed at keeping the supply and demand for money at equilibrium, as measured by growth in productivity and demand. Friedman focused on price stability, which is the equilibrium between supply and demand for money.

The rise of monetarism accelerated from Milton Friedman's 1956 restatement of the quantity theory of money. Friedman argued that the money could be described as depending on a small number of economic variables. Thus, where the money supply expanded, people would not simply wish to hold the extra money in idle money balances; *i.e.*, if they were in equilibrium before the increase, they were already holding money balances to suit their requirements, and thus after the increase they would have money balances surplus to their requirements. These excess money balances would therefore be spent and hence aggregate demand would rise. Similarly, if the money supply were reduced people would want to replenish their holdings of money by reducing their spending. In this, Friedman challenged a simplification attributed to Keynes suggesting that "money does not matter." Thus the word 'monetarist' was coined. Milton Friedman and Anna Schwartz in their book A Monetary History of the United States, 1867-1960 argued that the Great Depression of 1930 was caused by a massive contraction of the money supply and not by the lack of investment Keynes had argued. They also maintained that post-war inflation was caused by an over-expansion of the money supply. Monetarists of the Milton Friedman school of thought believed in the 1970s and 1980s that the growth of the money supply should be based on certain formulations related to economic growth. As such, they can be regarded as advocates of a monetary policy based on a "quantity of money" target. There was debate between Monetarists and Keynesians in the 1960s over the role of government in stabilizing the economy. Both Monetarists and Keynesians are in agreement over the fact that issues such as business cycles, unemployment, inflation are caused by inadequate demand, and need to be addressed, but they had fundamentally different perspectives on the capacity of the economy to find its own equilibrium and as a consequence the degree of government intervention that is required to create equilibrium. Keynesians emphasized the use of discretionary fiscal policy and monetary policy, while monetarists argued the primacy of monetary policy, and that it should be rules based. The debate was largely resolved in the 1980s. Since then, economists have largely agreed that central banks should bear the primary responsibility for stabilizing the economy, and that monetary policy should largely follow the Taylor rule – which many economists credit with the Great Moderation. The Global Financial Crisis, however, has convinced many economists and governments of the need for fiscal interventions and highlighted the difficulty in stimulating economies through monetary policy alone during a liquidity trap.

11.2. Rational Expectations School

Rational expectations is a hypothesis in economics which states that agents' predictions of the future value of economically relevant variables are not systematically wrong in that all errors are random. Equivalently, this is to say that agents' expectations equal true statistical expected values. An alternative formulation is that rational expectations are model-consistent expectations, in that the agents inside the model assume the model's predictions are valid. The rational expectations assumption is used in many contemporary macroeconomic models, game theory and applications of rational choice theory. Since most macroeconomic models today study decisions over many periods, the expectations of workers, consumers and firms about future economic conditions are an essential part of the model. How to model these expectations has long been controversial, and it is well known that the macroeconomic predictions of the model may differ depending on the assumptions made about expectations To assume rational expectations is to assume that agents' expectations may be individually wrong, but are correct *on average*. In other words, although the future is not fully predictable, agents' expectations are assumed not to be systematically biased and use all relevant information in forming expectations of economic variables. This way of modeling expectations was originally proposed by John F. Muth and later became influential when it was used by Robert Lucas, Jr. and others. Modeling expectations is crucial in all models which study how a large number of individuals, firms and organizations make choices under uncertainty. For example, negotiations between workers and firms will be influenced by the expected level of inflation, and the value of a share of stock is dependent on the expected future income from that stock.

Rational expectations theory defines this kind of expectations as being identical to the best guess of the future (the optimal forecast) that uses all available information. Thus, it is assumed that outcomes that are being forecast do not differ systematically from the market equilibrium results. As a result, rational expectations do not differ systematically or predictably from equilibrium results. That is, it assumes that people do not make systematic errors when predicting the future, and deviations from perfect foresight are only random. In an economic model, this is typically modeled by assuming that the expected value of a variable is equal to the expected value predicted by the model. For example, suppose that P is the equilibrium price in a simple market, determined by supply and demand. The theory of rational expectations says that the actual price will only deviate from the expectation if there is an 'information shock' caused by information unforeseeable at the time expectations were formed. Rational expectations theories were developed in response to perceived flaws in theories based on adaptive expectations. Under adaptive expectations, expectations of the future value of an economic variable are based on past values. For example, people would be assumed to predict inflation by looking at inflation last year and in previous years. Under adaptive expectations, if the economy suffers from constantly rising inflation rates (perhaps due to government policies), people would be assumed to always underestimate inflation. This may be regarded as unrealistic - surely rational individuals would sooner or later realize the trend and take it into account in forming their expectations. The hypothesis of rational expectations addresses this criticism by assuming that individuals take all available information into account in forming expectations. Though expectations may turn out incorrect, they will not deviate systematically from the expected values. The rational expectations hypothesis has been used to support some radical conclusions about economic policy-making. An example is the Proposition developed by Thomas Sargent and Neil Wallace. If the Federal Reserve attempts to lower unemployment through expansionary monetary policy economic agents will anticipate the effects of the change of policy and raise their expectations of future inflation accordingly. This in turn will counteract the expansionary effect of the increased money supply. All that the government can do is raise the inflation rate, not employment. This is a distinctly New Classical outcome. During the 1970s rational expectations appeared to have made previous macroeconomic theory largely obsolete, which culminated with the Lucas critique. However, rational expectations theory has been widely adopted throughout modern macroeconomics as a modeling assumption thanks to the work of New Keynesians such as Stanley Fischer. Rational expectations theory is the basis for the efficient market hypothesis (efficient market theory). If a security's price does not reflect all the information about it, then there exist "unexploited profit opportunities": someone can buy (or sell) the security to make a profit, thus driving the price toward equilibrium. In the strongest versions of these theories, where all profit opportunities *have been* exploited, all prices in financial markets are correct and reflect market fundamentals (such as future streams of profits and dividends). Each financial investment is as good as any other, while a security's price reflects all information about its intrinsic value.

11.2.1. Criticisms of Rational Expectations School

The models of Muth and Lucas (and the strongest version of the efficient-market hypothesis) assume that at any specific time, a market or the economy has only one equilibrium (which was determined ahead of time), so that people form their expectations around this unique equilibrium. Muth's math (sketched above) assumed that P^* was unique. Lucas assumed that equilibrium corresponded to a unique "full employment" level (potential output) – corresponding to a unique NAIRU or natural rate of unemployment. If there is more than one possible equilibrium at any time then the more interesting implications of the theory of rational expectations do not apply. In fact, expectations would determine the nature of the equilibrium attained, reversing the line of causation posited by rational expectations theorists. A further problem relates to the application of the rational expectations hypothesis to aggregate behaviour. It is well known that assumptions about individual behavior do not carry over to aggregate behavior. The same holds true for rationality assumptions: Even if all individuals have rational expectations, the representative household describing these behaviors may exhibit behavior that does not satisfy rationality assumptions (Janssen, 1993). Hence the rational expectations hypothesis, as applied to the representative household, is unrelated to the presence or absence of rational expectations on the micro level and lacks, in this sense, a microeconomic foundation. It can be argued that it is difficult to apply the standard efficient market hypothesis (efficient market theory) to understand the stock market bubble that ended in 2000 and collapsed thereafter; however, advocates of rational expectations say that the problem of ascertaining all the pertinent effects of the stock-market crash is a great challenge. Those studying financial markets similarly apply the efficient markets hypothesis but keep the existence of exceptions in mind.

Supply-side economics developed during the 1970s in response to Keynesian economic policy, and in particular the failure of demand management to stabilize Western economies during the stagflation of the 1970s, in the wake of the oil crisis in 1973. It drew on a range of non-Keynesian economic thought, particularly the Chicago School and Neoclassical School. Supply-side economics is a school of macroeconomics that argues that economic growth can be most effectively created by lowering barriers for people to produce (supply) goods and services, such as lowering income tax and capital gains tax rates, and by allowing greater flexibility by reducing regulation. According to supply-side economics, consumers will then benefit from a greater supply of goods and services at lower prices. Typical policy recommendations of supply-side economists are lower marginal tax rates and less regulation.

The Laffer curve embodies a tenet of supply side economics: those government tax revenues from a specific tax are the same (nil) at 100% tax rates as at 0% tax rates respectively. The tax rate that achieves optimum, or highest government revenues is somewhere in between these two values. However what most separates supply-side economics as a modern phenomenon is its argument in favor of a low tax rate. Thus, Supply-side economists argued for tax cuts for increasing supply and economic growth. As in classical economics, supply-side economics proposed that production or supply is the key to economic prosperity and that consumption or demand is merely a secondary consequence.

New Keynesian Economics Is A School of Contemporary macroeconomics that strives to provide microeconomic foundations for Keynesian economics. It developed partly as a response to criticisms of Keynesian macroeconomics by adherents of New Classical macroeconomics. Two main assumptions define the New Keynesian approach to macroeconomics. Like the New Classical approach, New Keynesian macroeconomic analysis usually assumes that households and firms have rational expectations. But the two schools differ in that New Keynesian analysis usually assumes a variety of market failures. In particular, New Keynesians assume that there is imperfect competition in price and wage setting to help explain why prices and wages can become "sticky", which means they do not adjust instantaneously to changes in economic conditions. Wage and price stickiness, and the other market failures present in New Keynesian models, imply that the economy may fail to attain full employment. Therefore, New Keynesians argue that macroeconomic stabilization by the government (using fiscal policy) or by the central bank (using monetary policy) can lead to a more efficient macroeconomic outcome than a laissez faire policy would.

Neo-liberalism is a political philosophy whose advocate support economic liberalization, free trade open markets, privatisation, deregulation, and decreasing the size of the public sector while increasing the role of the private sector in modern society. The term was introduced in the late 1930s by European liberal scholars to promote a new form of liberalism after interest in liberalism had declined in Europe. In the decades that followed, neoliberal theory tended to be at variance with the more laissez-faire doctrine of classical liberalism and promoted instead a market economy under the guidance and rules of a strong state, a model which came to be known as the social. In the sixties, usage of the term "neoliberal" heavily declined. When the term was reintroduced in the following decades, the meaning had shifted. The term neoliberal is now normally associated with laissez-faire economic policies, and is used mainly by those who are critical of legislative market reform. The term "neoliberalism" was originally coined in 1938 by the German scholar Alexander Rustow. It is defined as "the priority of the price mechanism, the free enterprise, the system of competition and a strong and impartial state." To be "neoliberal" meant that – in the name of liberalism – a modern economic policy is required. Neo liberalism was not a monolithic theory. At the outset it drew on different academic approaches such as the the Austrian School, the Chicago School of Economics etc. The term neo liberalism is almost never defined but used in several senses to describe ideology, economic theory, development theory, or economic reform policy. And it now suggests a market fundamentalism closer to the laissez-faire principles. This leaves some controversy as to the precise meaning of the term and its usefulness as a descriptor in the social sciences, especially as the number of different kinds of market economies have proliferated in recent years. Now a day the most common use of the term neoliberalism refers to economic reform policies such as "eliminating price controls, deregulating capital markets and lowering trade barriers", and reducing state influence on the economy especially by privatization and fiscal austerity. The term is used in several senses: as a development model it refers to the rejection of structuralist economics in favor of the Washington Consensus; as an ideology the term is used to denote a conception of freedom as an overarching social value associated with reducing state functions to those of a minimal state; and finally as an academic paradigm the term is closely related to neoclassical economic theory. Some people believe that the term is used as a pejorative for policies that deregulate the private sector and increase its role in the economy.

11.2.2. Dependence Theory

Dependency theory originates with two papers published in 1949 – one by Hans Singer, one by Raul Prebisch – in which the authors observe that the terms of trade for underdeveloped countries relative to the developed countries had deteriorated over time: the underdeveloped countries were able to purchase fewer and fewer manufactured goods from the developed countries in exchange for a given quantity of their raw materials exports. Dependency theory is a body of social science theories predicated on the notion that resources flow from a "periphery" of poor and underdeveloped states to a "core" of wealthy states, enriching the latter at the expense of the former. It is a central contention of dependency theory that poor states are impoverished and rich ones enriched by the way poor states are integrated into the "world system." The theory arose as a reaction to modernization theory, an earlier theory of development which held that all societies progress through similar stages of development, that today's underdeveloped areas are thus in a similar situation to that of today's developed areas at some time in the past, and that therefore the task in helping the underdeveloped areas out of poverty is to accelerate them along this supposed common path of development, by various means such as investment, technology transfers, and closer integration into the world market. Dependency theory rejected this view, arguing that underdeveloped countries are not merely primitive versions of developed countries, but has unique features and structures of their own; and, importantly, is in the situation of being the weaker members in a world market economy.

The premises of dependency theory are that: Poor nations provide natural resources, cheap labour, a destination for obsolete technology, and markets for developed nations, without which the latter could not have the standard of living they enjoy. Wealthy nations actively perpetuate a state of dependence by various means. This influence may be multifaceted, involving economics, media control, politics, banking and finance, education, culture, sport, and all aspects of human resource development (including recruitment and training of workers). Wealthy nations actively counter attempts by dependent nations to resist their influences by means of economic sanctions and/or the use of military force. Dependency theory states that the poverty of the countries in the periphery is not because they are not integrated into the world system, or not 'fully' integrated as is often argued by free market economists, but because of how they are integrated into the system. This introduces a paradoxical effect, in that although both the first and third-world countries are benefitting, the poorer side is being locked into detrimental economic position. They rely on the rich for the little work that is available to them, yet this causes a barrier from the nation growing independently. In a future perspective, such nations have no opportunity to improve their quality of life.

11.3. Indian Economic Thought

Indian economic thought is relatively little known either in India or elsewhere. The study of ancient Indian economic ideas provides a deeper insight into India's culture, tradition, and inherent national characteristics. The major sources of information about the economic ideas of Indian writers are Vedas, Arthasastra, the Ramayana and Mahabharata, Manusmriti, Sukraniti, and several other ancient Indian texts. Many scholars, particularly during the first half of the 20th century, attempted to make an objective and unbiased evaluation of ancient Indian economic ideas, explicitly or implicitly, contained in ancient Indian writings. Indian economic thought has certain special features which necessitates a separate study of Indian economic thought. The peculiar features of Indian economic ideas are:

1. **Descriptive:** The economic ideas are descriptive and not analytical. Indian thinkers paid more attention to practical problems and hence are realistic in nature.
2. **Mixed with politics:** Economic ideas were mixed with politics and were influenced by political factors.
3. **Less Emphasis to Material wealth:** Indian thinkers mixed ethics and economics together and did not give much importance to material wealth and welfare. They gave importance to moral aspects of life also. Thus the utility of the study to Indian economic thought is undoubtedly very great. It provides an economic interpretation of our historical past; and enables us to interpret the motives of economic activities of our ancestors. We may

classify Indian economic thought mainly into two as ancient economic thought and modern economic thought.

11.3.1. Ancient Indian Economic Thought

The history of Indian economic thought provides rich insights into both economic issues and the workings of the Indian thinkers. A study of the history of Indian economic thought provides the first overview of economic thought in the sub-continent. The sources of information available for the study of ancient Indian economic thought are Vedas, the Upanishads, the Epics - Ramayana and Mahabharata, Smritis and Niti Shastras particularly those of Manu and Shukra. Among these, the two most well-known ancient Indian writings are Arthasastra and Manusmriti. Kautilya was the important thinker, whose 'Arthasastra' has been considered the most reliable work on ancient Indian economic thought. It should be mentioned that ancient Indian thinkers had no clear conception of economics and their ideas were mixed with politics, ethics and economics.

Kautilya (350 BC-275 BC)

Kautilya was a professor of political science at the Takshashila University of ancient India, and later the chief Minister of Chandragupta Maurya who ruled over the mighty Indian empire in the 4th century BC. He wrote the Arthashastra ("Science of Material Gain" or"Science of political economy" in Sanskrit). Both names that are traditionally identified with Kautilya. He is known by the name of Kautilya because he was an expert in diplomacy and political strategy. Arthashastra deals extensively with aspects of political, economic, and social management and is the oldest book on management in the world. Arthashastra provides an authentic account of the economic and political ideas of the day. It also contains information about law, agriculture etc. prevailed at that time. It also deals in detail with the qualities and disciplines required for a wise and virtuous king. According to Kautilya, the king's happiness lies in the happiness of his subjects and his welfare lies in the welfare his subjects. It was by Kautilya around 350 BC. Kautilya has given vivid description of the administrative procedures, the duties of kings, ministers and government officers, in his book, Arthasastra. It covers almost every aspect of the theory and practice of economics. It also deals with nature and purpose of material wealth, varta - agriculture and animal husbandry, dignity of labour, trade, public finance, population, slavery, welfare state, social security, interest, price control, socio-economic institutions and town planning. His book contains ample ideas on a welfare state. We may briefly summarise the economic ideas prevailed in ancient India as elaborated by Kautilya.

Nature and Purpose of Material Wealth

Kautilya says that artha (sound economies) is the most important thing and dharma and karma are dependent on it. Hence the king shall be ever active in the management of the economy. The root of wealth is (economic) activity and lack of it results in material distress. In the absence of (fruitful economic) activity, both current prosperity and future growth will be destroyed. A king can achieve the desired objectives and abundance of riches by undertaking (productive) economic activity.

The term Artha (wealth) occupied an important place as one of the Purusharthas, in the system of economic thought during the Vedic Period. Artha or the material wealth or prosperity was essential for the stability and maintenance of the social structure and organization which provide means and opportunities to a decent living. The most common word used at that time as Varta meaning the national economy. The national economy consists of agriculture, animal husbandry and trade. Later on money lending and artisanship were also brought under Varta. The King was expected to have a good grasp of Varta or Economics. The term Arthasastra was wider in scope than Vartha and it was a combination of economics, political science and jurisprudence. Asthasatra is that science which describes the actions and administration of Kings in accordance with the dictates of revelation and of law as well as the means of appropriate livelihood. Arth or material prosperity was necessary for the smooth functioning of social structure, organization and institutions.

Ancient Indian scholars laid considerable emphasis on agriculture including animal husbandry. It was regarded as the basic source of new wealth. Agriculture was the highest occupation in society. Agriculture is the most important constituent of the economy. Three principal vocations are recognized as providing men with the means of livelihood namely, krsi (agriculture), pasupalya (cattle rearing) and vanijya (trade). The three together constitute varta (derived vritti,-livelihood). In those days the state and the community were responsible for the development of agriculture for which waste land were to be cultivated. The lands which were neglected by absentee landlords were to be taken away and given to those who could cultivate them with greater advantage. With respect to taxes on agriculture, avoid extremes of either complete absence of taxes or exorbitant taxation. The economic thinkers of ancient India regarded that labour as unproductive which failed to achieve its end. They never said that agriculture was more productive than other types of labour. Kautilya clearly mentioned that women should help men in productive activities in agriculture and trade. Gold or bullion was regarded as a means of producing wealth; and trade was the source of industrialized capital. There was free trade in those days in India. Tolls, duties and customs were realized for revenue purposes. The state had framed trade regulations which show that commerce in ancient India had reached an advanced stage.

In ancient India there were no fear of large population as population could not grow beyond a reasonable limit because of constant wars between small states, and loss of life due to the lack of medical facilities. The state was to promote the welfare of people by regulating the life of people. It had to give subsidies for the development of trade, agriculture, irrigation, mines, cattle welfare, etc. Various regulations relating to financial transactions, weights and measures, essential industries, currency and exchange, public works, prevention of adulteration, usury, etc point towards the ideal of a welfare state. Good governance in Kautilya's literature is aimed at fulfilling the welfare of the people. "In the happiness of the King's subjects lies his happiness, in their welfare, his welfare. Whatever pleases him personally, he shall not consider as good, but whatever makes his subjects happy, he shall consider good."Our ancient economic thinkers supported the institution of private property. Individual and family can own land and the right in land was tansferable and saleable. The state dominated all forms of property and levied cesses and fines when needed. Interest was justified in ancient Indian the ground that capital was essentially productive. Interest in those days was part of profit. Kautilya proposed the public regulation of interest.

The starting point of the ideals of consumption in ancient India was the acceptance of the doctrine of the four ends of life. They are Dharma, Artha, Kama and Moksha. In those days four agents of production – land, labour, capital and organization appear to have been recognised. Land was regarded as the source of all wealth. Our ancient economic thinkers had a clear idea about the functions of the state. In the field of production, the government followed the principle of full freedom and enterprise to individuals within limits. It is important to note that private individuals could also undertake the manufacture and sale of commodities monopolized by the state. Taxation was regarded as one of the most important source of State revenue. Ancient Indian thinkers supported tax for beneficial purposes and not to be wasted by the government. The two principles that were followed in connection with the realization of taxes were: (i) It should be levied once a year and should not prove burdensome and (ii) Taxes should be levied according to the ability to pay. Kautilya's discussion of taxation has several underlying principles - the taxing power of the state should be limited, tax should not be felt to be heavy or excessive, tax hikes should be introduced gradually, tax should be levied in the proper place, time and form, and tax level should be equitable and reasonable. Ideally, the government should collect taxes like a honeybee that sucks just the right amount of honey from the flower so that both can survive. Kautilya's scheme of taxation involved the elements of sacrifice by the taxpayer, direct benefit to the taxpayers, redistribution of income (the state took care of the poor), and tax incentives for desired investments. Kautilya suggested forced loans formeeting deficit budgets.

Town planning which included the re-orientation of main roads and street, and the sub-division of city areas was found in a much developed form, particularly during the days of Mauryan Kings. The metropolitan city was established after a detailed and careful planning; and due emphasis was laid on the maintenance of sanitation and prevention of fire. Regarding social security, Kautilya emphasized that it was the prime duty of the state to protect the weak and the aged, to provide jobs to the unemployed, and to set up poor houses and charitable institutions. The study of ancient Indian economic thought leads one to the conclusion that the economic ideas were scattered and discontinuous. Economic life was governed by religious ideals and moral sanctions. People generally believed in the Karma theory. Economics was not regarded as a separate discipline but combined with ethics, politics and philosophy. State intervention was considered necessary for the protection of consumers and weaker sections of the community. Due emphasis was laid on the establishment of a welfare state. It was the duty of the King to maintain law and order and to regulate the life of people. The Kind should also adopt measures to increase their prosperity. Self – dependent village units provided economic stability in ancient India. The economic ideas of Kautilya are undoubtedly the precious gems of ancient economic thought.

Thiruvalluvar was a celebrated Tamil poet and philosopher and ancient Indian economic thought would be incomplete without the economic ideas of Thiruvalluvar. His celebrated work Thirukkural is one of the important works of the Sangam era and it is a work on ethics. It deals with the fundamentals of life. Valluvar's teachings were an appeal to the people and he was not a mere scholar but a prophetto his countrymen. Thiruvalluvar is thought to have lived sometime between the 2nd century BC and the 8th century AD. The basic economic ideas of Thiruvalluvar are contained in the four chapters of Thirukkural and it is one of the most revered ancient works in the Tamil language. It is considered a 'common creed', providing a guide for human morals and betterment in life. Thirukkural is a combined word formed by joining the two words Thiru (meaning revered) and Kural (is a form of poem writing style, like Ballad in English Poems). Thirukkural is divided into three sections: section one deals with Aram, good ethical behavior with conscience and honor ("right conduct"); section two discusses Porul, the right manner of conducting worldly affairs; and section three deals on Inbam, love between men and women. Although two sections, Aram and Inbam, are devoted to the private life of an individual more than half the couplets in Thirukural are grouped under Porul which discusses ethics in public life. Thus Thiruvalluvar gives more importance to righteous living in public life. According to Thiruvalluvar, the four principles for a prosperous society are (i) faith in God (ii) economic resources (iii) spiritual leadership and (iv) observance of moral law. The economic ideas of Thiruvalluvar are found in the second part of the 'Kural' which is named as Porutpal (dealing with wealth).

Porutpal literally means things, wealth etc. and hence include all material things than can be acquired which a man requires in daily life for maintaining ones family. According to Valluvar, wealth is only a means and not an end. He was against hoarding of wealth and to him a hoarder is a burden to the earth. Thus, Porutpal covers all consumer and producer goods. According to Valluvar, a society should have freedom from hunger, freedom from disease and freedom from fear of external and internal aggression.Thiruvalluvar was aware of the dangers of poverty and hence cursed poverty and he was against begging. Hence to him poverty and begging are the greatest social curses Valluvar attached very much importance to agriculture. He favoured peasant proprietorship and was against absentee landlordism. According to him all are dependent on agriculture for food.

According to him, there are four functions to the state and they are the creation of revenues, collection of revenues, management of revenues and public expenditure. He was not in favour of compulsion in taxation and preferred balanced budgeting. He was of the view that the king should promote education and health of the people. He believed that people should be free from hunger which increases welfare of the society. Thence, he was an advocate of welfare state.When we compare Kautilya and Thiruvalluvar, we find the following differences between them in their economic ideas.

I. Thiruvalluvar attached importance to agriculture whereas Kautilya attached importance to wealth.

II. Thiruvalluvar advocated taxation by consent whereas Kautilya advocated compulsory taxes to increase states revenue for the benefit of the society.

III. According to Kautilya the economy is based on caste system while according to Thiruvalluvar, the economy is not based on caste system.

IV. We can see an ethical foundation of society in Thiruvalluvar whereas in Kautilya's Arthasastra, ethical principles are absent. We may conclude that Valluvar was a great sage and poet and was not an economic theorist. He tried to synthesise the principles of ethics, economics and politics in his Thirukkural.

12
Modern Indian Economic Thought

12.1. Modern Indian Economic Thought

Economic thought in modern India was largely influenced by the British rule in India. The British rule has impoverished the country and according to Indian thinkers this has been the major reason for the abject poverty of Indians. Hence, Indian nationalism and widespread poverty of the masses in India influenced the economic thought of modern India. We may briefly explain the economic ideas of Indian economists in the following paragraphs. It should be mentioned that the economic ideas of Indian writers are mixed with ethics and politics.

(a) Dadabhai Naoroji (1825 – 1917)

Dadabhai Naoroji, the Grand Old Man of India, was the father of Indian nationalism. He was from a rich Parsi family of Bombay and was an intellectual, educator, and a political and social leader. He was a Member of Parliament (MP) in the United Kingdom (House of Commons between 1892 and 1895), and the first Asian to be a British MP. Naoroji is also credited with the founding of the Indian National Congress, along with A.O. Hume. He was also member of Second International along with Kautsky and Plekhanov. At the early age of 25, he was appointed leading Professor at the Elphinstone Institution in 1850, becoming the first Indian to hold such an academic position. In 1854, he also founded a fortnightly publication, the Rast Goftar (or The Truth Teller), to clarify Zoroastrian concepts. By 1855 he was Professor of Mathematics and Natural Philosophy in Mumbai. Later, he became Professor of Gujarati at University College London. In 1874, he became Prime Minister of Baroda and was a member of the Legislative Council of Mumbai (1885–88). Naoroji was elected President of the Congress in 1886. He died in Bombay on 30 June 1917, at the age of 91. He was the first Indian to estimate the national income of India. Naoroji published Poverty and un-British Rule in India in 1901. His book brought attention to the draining of India's wealth into Britain.

(b) Economic Ideas of Naoroji

The economic ideas of Naoroji are found in his celebrated work: Poverty and Un-British Rule in India". He considered poverty as the major problem of India and attributed the mass poverty of India to the British rule. According to him, the continuous exploitation of India by the British government and the consequent draining of Indian wealth to Britain was the reason for the poverty of Indians. He estimated the per capita income in Bombay Presidency for the years 1867-70. He found that it was Rs.20 and reflected the poverty of the people as he found that at least Rs.34 is needed to satisfy the minimum needs of life.

(c) The Drain Theory

Dadabhai Naoroji is known for drain theory by which he focused on the drain of India's wealth to England through the colonial rule. The consequence of foreign domination was the drain of wealth of India to Britain. Through his work with economics, Naoroji sought to prove that Britain was draining money out of India. Naoroji described 6 factors which resulted in the external drain. Firstly, India is governed by a foreign government. Secondly, India does not attract immigrants which bring labour and capital for economic growth. Thirdly, India pays for Britain's civil administrations and occupational army. Fourthly, India bears the burden of empire building in and out of its borders. Fifthly, opening the country to free trade was actually a way to exploit India by offering highly paid jobs to foreign personnel. Lastly, the principal income-earners would buy outside of India or leave with the money as they were mostly foreign personnel. The drain of wealth took place in a number of ways. First large remittances were made by the British officials of their savings in India. Second large remittances were made in the form of salaries and pensions. Third, India often had to pay for government expenditure in England and of course in India too. Fourth, non-official Britishers made remittances from their business profits in India. The money which has gone out of India to England has once again come back as British Capital has monopolized trade and industry. It has once again resulted in the drain of wealth. Thus, the drain has become continuous and it has affected capital formation in India. Naoroji estimated that the drain was about 3 million pounds at the beginning of 19th century to 30 million pounds towards the end of the 20th Century. When referring to the Drain, Naoroji stated that he believed some tribute was necessary as payment for the services that England brought to India such as the railways. However the money from these services was drained out of India. India was paying tribute for something that was not bringing profit to the country directly. Instead of paying off foreign which other countries did, India was paying for services rendered despite the operation of the railway were already profitable for Britain. This type of drain was experienced in different ways as well, for instance, British workers earning wages that were not equal with the work that they have done in India, or trade that undervalued India's goods and overvalued outside goods. Englishmen were encouraged to take on high paying jobs in India, and the British government allowed them to take a portion of their income back. Furthermore, the East India Company was purchasing Indian goods with money drained from India in order to export to Britain. Free trade allowed Britain to exploit India.

Naoroji believed that to solve the problem of the drain it was important to allow India to develop industries. Over time, Naoroji became more extreme in his comments as he began to lose patience with Britain.

According to Naoroji, certain measures are needed to reduce Indian poverty and they are:

I. India and England should pay all salaries to their people within their boundaries.

II. As the Englishmen were paid reasonable salaries while they served India, there was no need to pay pension to them.

III. As there was no danger of invasion of India by sea, India should not be charged any portion of the expenditure incurred for the Indian navy.

12.1.1. Gandhian Economics

M.K.Gandhi (1869-1948). Gandhian economics is a school of economic thought based on the ethical foundations founded by Indian leader Mohandas Gandhi. It is largely characterised by its affinity to the principles and objectives of nonviolent humanistic socialism, but with a rejection of violent class war and promotion of socio-economic harmony. Gandhi's economic ideas also aim to promote spiritual development andnharmony with a rejection of materialism. The term "Gandhian economics" was coined by J. C. Kumarappa, a close supporter of Gandhi. Thus, Gandhian economic thought is based on four fundamental principles- Truth, Non-violence, Dignity of Labour and Simplicity. Gandhi believed in simple living and high thinking. Gandhi opposed capitalism as it is based on the exploitation human labour. To him, moral progress is more important than material progress. Salient Features of Gandhi's economic Thought are:

(a) Village Republics and Sarvodaya

Gandhi was interested in developing the villages as self sufficient units. Gandhiji believed that real India was to be found in villages. He suggested the development of self-sufficient and self dependent villages. Village Swaraj was his ideal. Gandhi has given top priority to Khadi programme. He worked and advocated for the Sarvodaya or the welfare of all. He thought that the development of rural civilization on the basis of self-sufficient villages would result in the decentralization of political and economic power. Thus, he favoured Grama Swaraj or self-ruling villages. He opposed extensive use of machinery, urbanization and industrialization on the ground of destruction of villages and large scale unemployment in the villages.

(b) On Machinery

Gandhi opposed machines as described machinery as great sin. However, Gandhi was not against machinery in general. He was against labour saving machinery. He welcomed instruments and machines that saved labour and lightened the burden of millions of cottagers. Gandhi emphasized that he was against large scale production only of those things that villagers can produce without difficulty. He believed that the machinery was harmful when the same thing could be done easily by millions of hands not otherwise employed. In short, Gandhi was aware of the danger of technological unemployment. Gandhiji's ideas on machinery are still relevant as many decades of machine using planned development failed to solve the problem of unemployment in India.

(c) Industrialisation

Gandhi considered industrialization a curse to mankind and he was a sharp critic of large scale industrialization. He was aware of the evils of large scale industrialization. He described industrialization as a manifestation of capitalism. Instead he suggested the development of village industries as it will promote the welfare of all. Gandhi opposed large scale industrialisation on the grounds on social justice also. Machinery would enrich the few at the expense of the many. He believed that it would result in the concentration of economic power in the hands of a few. He believed that large scale industries will destroy the handicrafts in villages. Gandhi advocated state ownership of public utilities.

(d) Decentralisation

Non-violence was the basic principle of Gandhi's life. He advocated a decentralised economy. His decentralisation includes production at a larger number of places on a small scale. Thus Gandhi wanted production by masses and not mass production. He wanted the revival of cottage industries. He liked coordinated village communities.

(e) Bread Labour

Gandhi believed in the dignity of human labour. He believed that God created man to eat his bread by the sweat of his brow. By bread labour (body labour) Gandhi meant manual labour Gandhi preached and practiced the gospel of manual work.

(f) The Doctrine of Trusteeship

Gandhi has developed the doctrine of trusteeship to provide an alternative to Marxian socialism. According to the trusteeship doctrine, the rich man should consider him as a trustee of the society for all the wealth he accumulated. Hence, Gandhi desired the capitalists to become trustees of the nation by running their business efficiently for the welfare of the people. Thus, according to Gandhi, if one had inherited a big fortune (accumulated money), the entire amount did not belong to him, but to the entire community. One who had accumulated that wealth was entitled to only that part of it which was necessary for an honourable living. The rest of the accumulated wealth belonged to the entire society and must be spent on the welfare of all. The important features of the Trusteeship formula of Gandhi's ideal social order are:

I. Trusteeship provides a means of transforming the present capitalist order of society into an egalitarian one. It gives the present owning class the chance of reforming themselves.

II. It does not recognize any right of private ownership of property except in as much as it may be permitted by society for its welfare.

III. It does not exclude legislative regulation of the ownership and use the use of wealth.

IV. An individual will not be free to use his wealth for selfish satisfaction or in disregard of the interest of the society.

V. As there is a minimum living, there is a limit to the maximum income that could be allowed to a person in society.

VI. The character of production is to be determined by the social necessity and not by personal whim or greed.

(g On the Food Problem

Gandhi was against any sort of food controls. He believed that such controls will create artificial scarcity for food grains.

(h) On Population

He was against control of population by the use of contraceptives. He was in favour of birth control through brahmacharya or self control. He did not agree with the view that food shortage was due to the growth of population.

(j) On Prohibition

Gandhi advocated cent percent prohibition. He regarded the use of liquor as a disease rather than a vice. he felt that it was better for india to be poor than to have thousands of drunkards.

12.2. Relevance of Gandhian Economic Thought to India

There are economists who believe that Gandhian economics is utopian in nature. They regard Gandhi as a medieval mystic who tried to put back the clock of human progress. Gandhian Economics is based on ethical values and the dignity of man. Gandhian economics do not draw a distinction between economics and ethics. Economics that hurts the moral well-being of an individual or a nation is immoral, and therefore sinful. It is regarded as an alternative to western socialism. Sometimes it is said Gandhism is communism minus violence. Gandhism tries to bring communism without violence or with the consent of capitalists. Gandhi was a practical idealist. His economic thought is basically sound and is relevant to our times. His concept of trusteeship, decentralization of economic activities, labour intensive technology and priority to weaker sections are all relevant today. Gandhi has often advocated social justice and equality. According to him if mankind was to progress and to realize the ideals of equality and brotherhood, it must act on the principle of paying the highest attention to the prime needs of the weakest sections of the population. Gandhi was not against machinery as such. What he intended was that in a country like India where capital was scarce and labour abundant, it would be profitable to use labour intensive industries. He feared that use of machinery would result in technological unemployment. The mad race in industrialization and urbanization has resulted in the pollution of air and water. In the ideal social order of village economy, environmental pollution will not be a problem.

Dr. B.R. Ambedkar (1891-1956). Dr Bhim Rao Ambedkar celebrated as the "Father of Indian Constitution," found economics closest to his heart and got his doctorate for a thesis on "The Problem of the Rupee". Ambedkar was a keen student of economics and got his MA for his thesis on 'Ancient Indian Commerce' and M.Sc (London) for his thesis on 'The Evolution of Provincial Finance in British India'. His evidence before the Hilton-Young Commission was an important contribution to the discussion of currency problems in India. He gave expression to his thoughts on such issues as small-holdings, collective farming, land revenue and abolition of landlordism. It covered nearly four important decades – 1917 to 1956, and touched on all major political and economic events. He was a Professor of Economics in Mumbai's Sydenham College in the early 1930s. Ambedkar strongly believed that the fundamental cause of India's backward economy was the delay in changing the land system. The remedy was democratic collectivism that entailed economic efficiency, productivity and overhauling the village economy. This, he said, would wipe out elements of economic exploitation and social injustice. His idea of economic realism sought both freedom and welfare.The essential feature of his approach to economic problems was the condemnation of such extreme views as laissez-faire and scientific socialism. Mixed economy was the cornerstone of his economic ideas. He advocated an end to the glaring social and economic inequalities produced by the capitalist system. He realised that the solution to the problem of the untouchable landless labourers depended upon the solution to Indian agricultural problems. He focused on the injustice in basing the assessment of land revenue on income and advocated that land revenue be brought under the income tax. His work "The Problem of the Rupee" was considered an instructive treatise. He wrote that closing of the Mints would prevent inflation and disturbances in the internal price level. He advocated that the standard of value should be gold and the elasticity of currency should come from this source. Dr Ambedkar was conferred posthumously the Bharat Ratna, the nation's highest civilian honour, on his 99th birth anniversary in 1990.

12.2.1. Economic Ideas of Ambedkar

Ambedkar's economic philosophy is expressed in his ideas on mixed economy (socialism), state ownership of industries, industrialisation, etc. He belongs to group of liberal thinkers but by and large his orientation is that of a socialist. Strangely he does not agree with Karl Marx as his ideas and methods appeared violent to Ambedkar. His economic ideas reflected his concern for the downtrodden and his desire for their development and upliftment in the social ladder. In order to eradicate exploitation by landlords and capitalist he advised mixed economy or state socialism.

(a) Land Reform

Ambedkar's main concern was the nature of land holdings, especially their fragmentation and division into unproductive and economically in viable structure. According to him, consolidation may eliminate the evils of scattered holdings but not the evils of small holdings unless the consolidated holding was an economic holding. According to Ambedkar, the basic problem of Indian agriculture was that it was unable to generate surplus which ultimately led to scarcity of capital. This resulted in the inefficient use of resources, surplus labour and superfluous employment which by raising the cost of production culminated in low agricultural productivity. Thus the process of overall economic growth gets adversely affected. Ambedkar was of opinion that the process of consolidation of holding could eradicate the ill effect of uneconomic holding and usher in the progress of the cultivators. He favoured successful growth of industrialization. He stated "Industrialization facilitates consolidation. It is a barrier against future subdivision and consolidation." Ambedkar was the first person to raise his voice against Khoti system. The Khoti system was one of the minor land revenue tenures in the former Bombay presidency. It was found mostly in Ratnagiri district and in some parts of Kolaba and Thane districts. In Khoti system unlike Ryotwari system, the government employed the services of the Khot for the purpose of collection of revenue.Under this system the khot was obliged to collect revenue from the tenants and to pay a certain part of it to government. Once such a payment was made the Khot enjoyed unrestricted freedom to do whatever he liked to the tenants. On 17 September 1937 Ambedkar introduced in Bombay Legislative council a Bill seeking the abolition of the Khoti system. By introducing this bill Ambedkar sought to put an end to the exploitation of the actual tillers of the land and established a direct relationship between them and the government. Ambedkar was of the opinion that the state should acquire all the agricultural land held by private individuals like the owners, tenants or mortgages and pay them compensation equal to the value of land. The state should provide the land required into farms of standard size and should let out the farms for cultivation to the residents of the villages.

(b) Currency Problem

Ambedkar's writings on currency problems constituted in his book "The problem of the rupee" and his evidence before the Hilton Commission in 1926. He brought a new perception on the monetary issue. He was not in favour of linking the rupee with gold and recommended establishment of a fully managed inconvertible currency with fixed limit of issue. Ambedkar stated " It is much better to introduce a currency system which will do away with the exchange standard and also the gold standard reserve. According to Ambedkar the pure gold standard comprises use of gold in some form convertible standard, paper money is also issued in addition to gold coins and is pledged to be redeemable in gold. In contrast under the gold exchanged standard the medium of exchange comprises only paper money which is kept exchangeable at fixed rates with gold and authorities back it up with foreign currency reserves of such countries as on the gold standard. Ambedkar vehemently criticized Keynes and other supporters of the gold exchanged standard and argued in favour of the gold standard of a modified form. Ambedkar argued that the gold exchange standard does not have the stability of the gold standard. Under the gold standard, additions to the supply of currency are so small that stability is not affected. Ambedkar provided statistical evidence to show that prices varied much less under the gold standard than under the gold exchange standard in Indian context. Ambedkar favoured Gold currency and wanted to close down the mint as this will eradicate the money inflation and imbalances in internal payments. For the flexibility in currency he was of the opinion that Gold is a suitable measure for the flexibility of money.

(c) Decentralisation of Imperial Finance

Ambedkar's thesis the evolution of provincial finance in British India dealt with the history of public finance. He discussed in detail about the centre province financial relationship during 1833-1921. Ambedkar demonstrates how centralisation of government finances which prevailed in India during 1833 through 1871,was a failure on account of faulty fiscal system marked by injurious taxes and unproductive or extravagant expenditure. The system of Imperial Finance was started in India in 1833. It became so elaborated not e in 1858 when the crown took over from the East India Company that no province had any separate power for legislating any separate financial resources. The imperial government was responsible in law but did not administer the country whereas the provincial governments administered the country but had no responsibility in law. Under this system budgets were prepared by the provincial government but the responsibility for the finding ways and means for financing them rested on the government of India. There was practically no power of creating or modifying any appointments in service. The restriction involved provided the Government of India with an opportunity to interfere with all the details of provincial administration." These led to extravagant demands and since the government of India did not possess the necessary machinery to appraise the demands and to control the expenditure, it often had to yield. Government finances came under severe strain and it was realised that the provincial government must draw up their own revenue and expenditure budgets. Accordingly the regime of provincial budgets came into force in 1871. Provincial finance was introduced in the form of budget by assignment (as referred by Ambedkar) during 1871-72 through 1876- 77. Under this scheme financial responsibilities of certain departments of administration were delegated to the provinces and the receipts accruing from the imperial from those services were handed over to them with fixed lump sum assignment from the treasury. According to Ambedkar this scheme led to high taxes and made taxation more iniquitous. The next phase in the evolution of provincial finance was the budget by assigned revenue which was operational during the period 1877-78 through 1881-82. It gave the provinces certain sources the yield of which largely depends on good management for the growing needs of the provincial services.

The third phase in the process of evolution of provincial administration was the scheme of budget by shared revenues. According to Ambedkar the earlier schemes were deficient as the sources of revenues assigned to provinces had little room for enlargement over the time. But the new scheme of shared revenues specific revenues were collected either by the imperial government of provinces and a portion of the proceeds was allotted to the other. The is scheme was most successful and lasted for 38 years and in 1920-21 major fiscal reforms were introduced in 1921 which in the opinion of Ambedkar were not induced by any inherent defects in the system. Thus, Ambedkar dealt in detail with the mechanism of finance of the centre- provincial government under the old phase.

(c) State Management

Ambedkar emphasised on the nationalisation of economy. He was of the view that state should manage the economy that the production might reach the optimum level and the benefits must not be taken away by the capitalist. The benefits must be distributed equally. He stood for the progressive transformation of society, removing glaring social and economic inequalities that were due to the capitalist system. B.R. Ambedkar was a firm believer of socialism. According to him, "state socialism is essential; for India's industrialisation. Private economy cannot do so and if it makes an attempt it would give way to economic disparities, as it can be visualised in the case of Europe. His view on nationalization of industries and life insurance showed his great concern for the economic problems faced by India. The formation of the public sector and the establishment of Life Insurance Corporation showed that he was in agreement with other leading economists. He remarked that industrialisation of India was a necessary thing. But side by side the principle of state management and state ownership of industry must be adopted. Amenities like social insurance and control over employment, dismissal were vital to the progress of industry. About these measures the lower middle class must be protected. He firmly believed that by eliminating exploitation, the industrial harmony can be established through labour welfare and congenial industrial relation. He remarked "we have attained political freedom and equality but without economic and social equality this is quite insufficient". Ambedkar emphasised more on economic and social freedom and equality. his concept of society and socialism aimed for the welfare of the poor classes, ending inequality based on birth eliminating discriminatory practises in social behaviour patterns reorganising the political economy for the benefit of all maintaining full employment and education, providing social security for the weak and the sick.

(d) Labour Problem

Ambedkar supported trade union movement and right to strike against capitalism. He wanted participation of labourers in industrial management. He paid attention to joint councils employment exchanges and earned leave for permanent workers, welfare activities, conciliation, trade disputes. According to him industrial peace would prevail if it is based on social justice. He introduced 5% to 6% reservation for Schedule castes in Central government. He helped the untouchable students wishing to take technical education in foreign countries. In his inaugural address to the conference of the Regional Labour Commission Ambedkar stated "three things are necessary to mitigate or prevent industrial disorder – Machinery for Conciliation, amendment in the Trade Disputes Act and Minimum Wage Legislation. According to him, the industrial peace could be realised if it was based on social justice." Ambedkar, while discussing the labour problem, had in his mind not only industrial labour but also agricultural labour. According to him, similar condition of work provident fund employer's liability, workmen's compensation, health insurance including invalidity pensions should be open to all sorts of labour whether it was Industrial labour or agricultural labour.

According to Ambedkar the caste system in India was a major obstacle to economic growth and development. The caste system didn't allow people to teach their professional skills to any person belonging to other caste. Only the members of their own caste were allowed to learn the profession. Thus if a person had the skill necessary for a particular job he would not accept the profession of a caste lower than his own. In a dynamic industrial set up the individual must be free to choose his occupation. But due to social religious restriction on intern occupational mobility has following consequences: Firstly, by not permitting readjustment of occupation, caste become a direct cause of much of the unemployment in various groups, as a religious Hindu would prefer to be unemployed rather than getting employed in profession not assigned to his caste. Second, individual justice and economic efficiency demand that competition exists in factor market. Due to the restriction on inter-occupational mobility of labour, capital and entrepreneurship across caste groups the caste system creates segregation in each of these markets. Labour and capital thus does not flow from one occupation to another even if the wage rate or rates of return on investments are higher in the alternative occupations. This brings about a high level of inefficiency in resource allocation. The division of occupation is not based on individual choice. Some of the occupations are socially degrading, and people are forced in these occupations on account of their caste origin. Such people cannot derive job satisfaction. As economic organisation caste is therefore a harmful institution in as much as it involves the subordination of man's natural powers and inclination to the exigencies of social rules. This also results in disassociation of intelligence from work. The dignity of labour is nearly absent in the general scheme of theory of caste. The practise of untouchability has ruined the nation and the economy as a whole. Those belonged to the lower strata of society or the untouchables were not allowed to mix with others and denied basic civil right. They had no access to sources of income. They had no right to education. They had no right tom property and were prohibited from acquiring wealth. They could not use their labour for their own upliftment. The Caste system has crippled the untouchable as it denied them economic security and independence. In the opinion of Ambedkar the economics of the caste system had six type s of consequences: 1) it divided labour ii) it disassociated work from interest iii) it disconnected intelligence from the manual labourer iv) it devitalized persons by denying them right to cultivate vital interest v) it prevented mobilization vi) it deprived Sudras of all economic avenues of employment and took them nearly to the position of a slave.

Ambedkar characterised Gandhism as conservatism and purely imaginary. He remarked that Gandhism meant return to village and making the village self sufficient and as such it made Gandhism a mere matter of regionalism. Gandhiji vehemently opposed machinery. He idolized charkha (spinning wheel) and called western civilisation as the creation of satan. He even said "I would not weep over the disappearance of machinery or consider it a calamity". Ambedkar opposed

Gandhism because, in his opinion, Gandhism was against democracy and also against mechanisation and civilization which were vital for progress of a country. According to Ambedkar, democratic society cannot be indifferent to machinery. Gandhism might well be suited to a society which did not accept democracy as its ideal. He maintained that the ills were not due to machinery and they were wrong due to social organisation which has made private property and pursuit of personal gain matters of absolute sanctity. He remarked that "Gandhism is a paradox. It seeks to maintain intact a social structure which permits the domination of one class by another on a hereditary basis which means a perpetual domination of one class by another." Ambedkar didn't have faith in laissez faire or Trusteeship of Gandhiji. He believed in socialism and advocated state socialism to ensure proper working of national economy. He favoured land acquisition by the state for distribution among poor. In his view socialism will not come without directive principle. Ambedkar viewed, Gandhism, a philosophy of the well to do and the leisured class. It deluded people into accepting misfortunes by presenting them as best of good fortunes.

Ambedkar's views on communism are presented in his essay "Buddhism and Communism". Like Marx, Ambedkar believed that there is exploitation in the world, the poor are exploited by the rich; there is enslavement of the masses by the privileged few which leads to perpetuation of poverty and ultimately suffering. Ambedkar did not sympathise with the Marxist paradigm of development. Ambedkar was of the opinion that economic relationships are not everything in human life nor the economic motive was the sole driving power behind all human activity. He emphasised that exploitation has many dimensions like social, religious and political. In Ambedkar's opinion, communism advocates revolution, it recommends revolutionary methods for overcoming the opposition of the capitalists for establishment of the dictatorship of the proletariat and any action that serves the interest of the proletarian revolution is ethical and there is no consideration for human life. Ambedkar being an ardent follower of democracy and human right was against any form of dictatorship. He was not in favour of violent change rather he believed that constitutional provision and democratic means could bring desired reformation. The control of all means of production and the abolition of private property as advocated by Marx were not accepted by Ambedkar. Ambedkar was not in tune with Marxian convictions that abolition of private property would bring an end to the poverty and suffering of the have nots and that the state is a temporary institution which will wither away in course of time. Ambedkar believed that state is fundamental to human existence. It has many important roles to play. He maintained that state would continue to exist as long as human society survived.

He believed in classless society but not in a stateless society. He sought active involvement of the government in economic development. In his book, "States and Minorities", he stated that state has an obligation to plan the economic life of the people. Ambedkar's concept of state socialism is based on these tenets:

a) State ownership economy agricultural land and key industries
b) Maintenance of productive resources by the states
c) A just distribution of the common produce among different people without any discrimination of caste or creed.

Ambedkar was primarily an economist and his achievement in the field was primarily overshadowed by his achievement in his fields of politics. His political ideas had moral purpose. He was more concerned about the human welfare. He had unflinching faith in socialism and believed that it would solve the problem of the under privileged. Mixed economy was the cornerstone of his philosophy. He was concerned about the equitable distribution of wealth among people. He advocated the modernization of Indian society and economy. Ambedkar also linked salvation and emancipation of untouchable people with the emergence of modern urbanised and industrialised India with modern liberal ideology.

13
Prominent Indian Economists

13.1. Prominent Indian Economists

Mahadev Govind Ranade (1842 –1901). Mahadev Govind Ranade was a distinguished Indian scholar, social reformer and author. He was a founding member of the Indian National Congress and owned several designations as member of the Bombay legislative council, member of the finance committee at the centre, and the judge of Bombay High Court. A well known public figure, his personality as a calm and patient optimist would influence his attitude towards dealings with Britain as well as reform in India. During his life he helped establish the Poona Sarvajanik Sabha and the Prarthana Samaj. He published books on Indian economics and on Maratha history. His most important works are "Essays on Indian Political Economy" (1898) and The Rise of Maratha Power (1900). He saw the need for heavy industry for economic progress and believed in Western education as a vital element to the foundation of an Indian nation. He felt that by understanding the mutual problems of India and Britain both reform and independence could be achieved to the benefit of all and insisted that an independent India could only be stable after such reforms were made. Ranade founded the Poona Sarvajanik Sabha and later was one of the originators of the Indian National Congress. Ranade was a founder of the Social Conference movement, which he supported till his death, directing his social reform efforts against child marriage, the shaving of widows' heads, the heavy cost of marriages and other social functions, and the caste restrictions on traveling abroad, and he strenuously advocated widow remarriage and female education. He was one of the founders of the Widow Marriage Association in 1861. After his death, she continued his social and educational reform work.

13.1.1. Economic Ideas of Ranade

(a) Method of Economics

Ranade attacked the deductive method of the classical economists. He favoured inductive and historical method. He believed that economic laws are not universally applicable and they were relative. Ranade emphasized the importance of economic education in promoting the wealth of a nation.

(b) Role of the State

Ranade conceived a planned economy for India with the object of establishing a welfare state.He believed that the state should direct and channelize the activities of individuals and institutions,towards well determined goals. In India as the private initiative was shy, the state should undertake measures for the industrial development of the country.

(c) India's Poverty

Ranade believed that mass poverty had been there in India even before the advent of the British rule. However, he believed that poverty was intensified by the British rule. According to him,overdependence on agriculture, backward industries, inadequate credit by banks and land policy of the government were the reasons behind the poverty of the people.

(d) D.R. Gadgil (1901 –1971)

Dananjay Ramchandra Gadgil was an Indian economist and a former director of the Gokhale Institute of Politics and Economics and was deputy chairman, Planning Commission. He had very broad intellectual interests ranging from sociology, public affairs and history to biology and astronomy, and he had a personal library of over 3,000 books. Gadgil received a Master of Arts degree and a Master of Letters degree from the University of Cambridge. He held the post of vice-chancellor at the University of Poona. From 3 January 1966 to 31 August 1967, he was a nominated member of the Rajya Sabha. Healso served as president of the Indian Economic Association. He served on several committees including the All India Rural Credit Survey. He published 24 books, many articles and papers. His publications include 'The Industrial Evolution of India In Recent Times' (1924), 'Economic Policy and Development' (1955) and 'Planning and Economic Policy in India' (1961). He has analysed the industrial evolution of India and has contributed in the field of planning and finance. According to him, the most important events in the economic transition of the country were the decline of handicrafts and commercializationof agriculture. During the period from 1929 to 1939 he pointed out that the world depression and theemergence of provincial autonomy dominated the economic history of India. Dr. Gadgil sympathised with industrial workers and suggested higher wages and states measures for their welfare. He developed a formula in 1969 for distributing central assistance to states plans which is known as Gadgil formula. He suggested state regulation of prices on commodities like food grains, cloth, sugar, gur, edible oil, soap, tea, tobacco and kerosene for the benefit of rural population.

(e) C.N. Vakil (1895-1979)

Professor C.N.Vakil was the Director of the School of Economics and Sociology of the University of Bombay up to 1956.Thereafter; he worked as the Director of the UNESCO Research Centre at Calcutta up to 1960.His main contributions are in the field of fiscal policy, Indian Finance and Planning. His most important work is 'Financial Developments in Modern India - 1860-1924' (1929)' which was based on his research work carried out in London School of Economics. According to him our fiscal policy has not been for the benefit of the country and he reminded the Fiscal Commission that the interests of India alone should be the determining factor in decisions in future. According to him, the government expenditure is incurred is not conducive for the economic development of the country. In his book 'Economic Consequences of Divided India" he has analysed the more important economic consequences which arose with the partition of the country. Vakil along with P.R.Brahmananda developed the wage goods model of development.

(f) P.R. Brahmananda (1926-2003)

Professor P.R. Brahmananda was a noted economist and President of the International Economic Association (IEA). Prof. Brahmananda was the fourth Indian to be made the honorary president of the IEA. He guided 40 students for their doctorate at Bombay University including that of the current Governor of the Reserve Bank of India, Dr Bimal Jalan. A bachelor, Prof. Brahmananda was the son of P.R. Ramaiya. His mother, P.R. Jayalakshamma, was a former Deputy Mayor of Bangalore. Prof Brahmananda completed his doctorate from the School of Economics in Bombay and later worked as a research assistant to renowned economist Prof. C.N. Vakil. He was later a visiting professor of the Delhi School of Economics and the Director of the Bombay School of Economics. He was awarded the Rockefeller Foundation Fellowship in 1956, which could not be taken up. He was also a member of the panel of economists to the Planning Commission and Finance Ministry. He wrote more than 20 books and over 600 articles and papers to different economic journals and publications, with more than a dozen articles on development economics. In 1956, he published 'Planning for an Expanding Economy' with C.N. Vakil. In this book, they developed the 'Wage Goods Model' as an alternative to capital intensive strategy of Mahalanobis model of second Five Year Plan. According to the Wage Goods model employment has to be increased to reap higher growth rate of the economy by expanding the production of wage goods. Prof. Brahmananda's researches encompass the different aspects of general economic theory and he has pioneered studies in the subject of development economics in the light of the economic history and experience of a developing economy like India.

(g) Pranab Bardhan

Pranab Kumar Bardhan is Professor of Economics at the University of California, Berkeley (since 1977). Educated in Presidency College, Kolkata and Cambridge University, England, he had been on the faculty of MIT, Delhi School of Economics, and Indian Statistical Institute, before joining Berkeley. He has been Visiting Professor/ Fellow at London School of Economics, Trinity College, Cambridge, St. Catherine's College, Oxford, and University of Siena, Italy. He has done theoretical and field studies research on rural institutions in poor countries, on political economy of development policies, and on international trade and globalization. A part of his work is in the interdisciplinary area of economics, political science, and social anthropology. He has been on the editorial board of a number of economics journals, including The American Economic Review, the Journal of Economic Perspectives, the International Economic Review, and the Journal of Development Economics. He won a Guggenheim Fellowship in 1981 and the Mohalanobis Gold Medal of the Indian Econometric Society in 1980. He is the author of 12 books, more than 120 journal articles, and the editor of 12 other books. His research includes Development Economics, International Trade and Poverty.

(h) K.N. Raj (1924–2010)

Kakkadan Nandanath Raj was born in Thrissur District of Kerala. He is popularly known as K. N. Raj, an Indian economist who played an important role in India's planned development. He drafted sections of India's first Five Year Plan, specifically the introductory chapter when he was only 26 years old. He took B.A. from the prestigious Madras Christian College. He was a disciple of a well-distinguished economist Malcom Adiseshiah at Madras Christian College. His teacher pressed for him to go for higher studies in London School of Economics. His thesis was on the monetary policy of India's central bank. Raj was a companion of distinguished economists like Manmohan Singh, Amartya Sen and Jagdish Bhagwati. He was a veteran economist in the Planning Commission and worked out a plan to raise India's rate of savings in the post-Second World War period when the country was in need of foreign aid. He computed India's Balance of Payments for the first time for the Reserve Bank of India. Raj was an advisor to several prime ministers from Jawaharlal Nehru to P.V. Narasimha Rao. Dr. Raj was a Keynesian economist. He studied the application of Keynesian monetary theory in Indian context. Although he was a staunch leftist, he was a critical of Lenin's ideas of the State. Like all other Leftists, he opposed to the economic liberalisation in India. Raj joined Delhi University, where he was Professor of Economics and also Vice- Chancellor (from October 1969 to December 1970), spending a total of 18 years there. During that time, he was instrumental in setting up the Delhi School of Economics (DSE). After returning to Kerala from Delhi in 1971, Raj set up the Centre for Development Studies at Thiruvananthapuram, an institution that soon acquired an international reputation for applied economics and social science research. The work that Raj and his colleagues did for the United Nations in the early days of the CDS, and published in 1976, helped shape the contours of what later came to be called the "Kerala model" of development - the co-existence of low per capita income and very high physical quality of life indicators. It was really a paradox between economic and social development.

(i) P.C. Mahalanobis (1893–1972)

Prasanta Chandra Mahalanobis Was an Indian Scientist and applied statistician. He is best remembered for the Mahalanobis distance, a statistical measure. He founded the Indian Statistical Institute, and contributed to the design of large-scale sample surveys. Mahalanobis belonged to a Bengali family. P. C. Mahalanobis grew up in a socially active family surrounded by intellectuals and reformers. Mahalanobis received his early schooling at the Brahmo Boys School in Calcutta graduating in 1908. He then joined the Presidency College, Calcutta and received a B.Sc. degree with honours in physics in 1912. He left for England in 1913 to join the University of London but joined in King's College, Cambridge. He interacted with the mathematical genius Srinivasa Ramanujan during the latter's time at Cambridge. He went back to England and was introduced to the journal Biometrika. This interested him so much that he bought a complete set and took them to India. He discovered the utility of statistics to problems in meteorology, anthropology and began working on it on his journey back to India.

Many colleagues of Mahalanobis took an interest in statistics and the group grew in the Statistical Laboratory located in his room at the Presidency College, Calcutta. A meeting was called on the 17 December 1931 and the meeting led to the establishment of the Indian Statistical Institute (ISI). The Institute was initially in the Physics Department of the Presidency College. In 1933, the journal Sankhya was founded along the lines of Karl Pearson's Biometrika. In 1959, the institute was declared as an institute of national importance and a deemed university. During the course of these studies he found a way of comparing and grouping populations using a multivariate distance measure. This measure, D2, which is now named after him as Mahalanobis distance, is independent of measurement scale. His most important contributions are related to large-scale sample surveys. He introduced the concept of pilot surveys and advocated the usefulness of sampling methods. Early surveys began between 1937 to 1944 and included topics such as consumer expenditure, tea-drinking habits, public opinion, crop acreage and plant disease. In later life, Mahalanobis was a member of the planning commission, contributed prominently to newly independent India's five-year plans starting from the second. In the second five-year plan he emphasised industrialization on the basis of a two sector model. The Mahalanobis model, was employed in the Second Five Year Plan, which worked towards the rapid industrialization of India. He received one of the highest civilian awards, the Padma Vibhushan from the Government of India for his contribution to science and services to the country.

Mahalanobis died on 28 June 1972, a day before his seventy-ninth birthday. Even at this age, he was still active doing research work and discharging his duties as the Secretary and Director of the Indian Statistical Institute and as the Honorary Statistical Advisor to the Cabinet of the Government of India. The government of India decided in 2006 to celebrate his birthday, 29 June, as National Statistical Day. Professor P.C. Mahalanobis' contribution to the subject of statistics is incredibly varied. Mahalanobis had held the position of Statistical Advisor to the Government of India and Member, Planning Commission, beside many other distinguished appointments. The formulation of the D^2-Statistics, derivation of its Properties and its application are some of his most profound contributions. He received the Weldon Medal from Oxford University in 1944 and was elected a Fellow of the Royal Society, London, in 1945 for his fundamental contributions to statistics, particularly in the area of large-scale sample surveys. As Chairman of the United Nations Sub-Commission on Statistical Sampling, a position to which he was appointed in 1947, he tirelessly advocated the use of sampling methods to be extended to all parts of the world. As the Chairman of the UN Statistical Commission, he also helped in the spread of extremely useful statistical ideas to other countries. As a pioneer to government planning, his background to a number of initiatives in de-centralised planning is meritorious, one of them being the setting up of the National Sample Survey (NSS) entrusted with key responsibilities for the design, collection and analysis of sample survey data.

(j) V.K.R.V. Rao (1908–1991)

V.K.R.V. Rao was a prominent Indian economist, politician, professor and educator. He was born on July 8, 1908 at Kancheepuram in Tamil Nadu to Kasturirangachari and Bharati Bai. He had his early schooling in Tindivanam and Madras (Chennai). He was a recipient of the Padma Vibhushan. He served as a Union Minister for the Education in 1971, elected as member for Bellary in 1967 and 1971. He was awarded a Ph.D. in 1937 from Gonville and Caius College, Cambridge. The title of his doctoral thesis was "The national income of British India, 1931-1932". He studied with John Maynard Keynes. Rao received many awards. Rao established three noted institutions in Social Science research in India: Delhi School of Economics, Institute of Economic Growth and the Institute for Social and Economic Change. He was also instrumental in establishing the Indian Council of Social Science Research, Agro-economic Centres and Population Research Centres. Notable among his works are: Taxation of Income in India (1931), An essay on India's National Income -1925-29 – (1936); The National Income of British India (1940); India and International Currency Plans (1945); India's National Income 1950-80 (1983), Food, Nutrition and Poverty (1982). He served as Planning Adviser Food Department (1945-46), Food and Economic Adviser, Government of India at Washington (1946-47); Director, Delhi School of Economics, Delhi (1948-57); Vice Chancellor, University of Delhi (1957-60); Director, Institute of Economic Growth, Delhi (1960-63); Member, Planning Commission (1963-66); Union Cabinet Minister for Transport and Shipping (1967-69); Union Cabinet Minister for Education & Youth Services (1969-71); Director, Institute for Social and Economic Change, Bangalore (1972-77); National Professor, Government of India (1985- 1990). In the international sphere he was one of the central forces behind the institution of the United Nations Development Program (UNDP) and the IDA. He was the first Indian to compute the national income of India.

(k) I.G. Patel (1924–2005)

I. G. Patel was the fourteenth Governor of the Reserve Bank of India from 1 December 1977 to 15 September 1982. Patel was an economist and served as Secretary in the Ministry of Finance and thereafter as Deputy Administrator at the UNDP. After retiring from the Reserve Bank, Patel became the Director of The London School of Economics and Political Science (1984–1990). He was bestowed the Padma Vibhushan award in 1991 for his furthering of the field of economic science. The Indian Rupee notes of 1000, 5000 and 10,000 denomination and the gold auctions were demonetized during his tenure.

Amartya Kumar Sen was born in West Bengal 1933. Sen is an Indian economist who was awarded the Nobel Prize in Economics 1998. He was awarded Nobel prize for his contributions to welfareeconomics and social choice theory, and for his interest in the problems of society's poorest members. Sen is best known for his work on the causes of famine, which led to the development of practical solutions for preventing shortages of food. He helped to create the United Nations Human Development Index. In 2012, he became the first non-U.S. citizen recipient of the National Humanities Medal. He is currently the Professor of Economics and Philosophy at Harvard University. Sen's books have been translated into more than thirty languages over a period of forty years. He is a trustee of Economists for Peace and Security. In 2006, *Time magazine* listed him under "60 years of Asian Heroes" and in 2010 included him in their "100 most influential persons in the world". *New Statesman* listed him in their 2010 edition of "World's 50 Most Influential People Who Matter". Rabindranath Tagore is said to have given Amartya Sen his name ("Amartya" meaning "immortal"). Sen began his high-school education at St Gregory's School in Dhaka in 1941, in modernday Bangladesh. His family came to India following the partition of the country in 1947. In India Sen studied at the Visva-Bharati University school and then at the Presidency College, Kolkata. In 1953, he moved to Trinity College, Cambridge. To Sen, then Cambridge was like a battlefield. There were major debates between supporters of Keynesian economics and the "neoclassical" economists skeptical of Keynes. Sen submitted his Ph.D thesis on "the choice of techniques" in 1959 under the supervision of Joan Robinson. Between 1960 and 1961, Sen was a visiting Professor at Massachusetts Institute of Technology. He was also a visiting Professor at UC-Berkeley, Stanford, and Cornell. He has taught economics also at the University of Calcutta and at the Delhi School of Economics. In 1972, he joined the London School of Economics as a Professor of Economics where he taught until 1977. From 1977 to 1986 he taught at the University of Oxford, where he was first a Professor of Economics at Nuffield

College, Oxford and then the Drummond Professor of Political Economy and a Fellow of All Souls College, Oxford. In 1986, he joined Harvard as the Thomas W. Lamont University Professor of Economics. In 1998 he was appointed as Master of Trinity College, Cambridge. In January 2004, Sen returned to Harvard. He has served as president of the Econometric Society (1984), the International Economic Association (1986–1989), the Indian Economic Association (1989) and the American Economic Association (1994). He has also served as President of the Development Studies Association (1980–1982) and is a Honorary Vice-President of the Royal Economic Society, which he has been since 1988.

He presently serves as Honorary Director of Center for Human and Economic Development Studies at Peking University in China and is also a board council member of the Prime Minister of India's Global Advisory Council of Overseas Indians. In 1981, Sen published *Poverty and Famines: An Essay on Entitlement and Deprivation* (1981), a book in which he argued that famine occurs not only from a lack of food, but from inequalities built into mechanisms for distributing food. Sen also argued that the Bengal famine was caused by an urban economic boom that raised food prices, thereby causing millions of rural workers to starve to death when their wages did not keep up. Sen's interest in famine stemmed from personal experience. As a nine-year-old boy, he witnessed the Bengal famine of 1943, in which three million people perished. He presents data that there was an adequate food supply in Bengal at the time, but particular groups of people including rural landless labourers and urban service providers like haircutters did not have the monetary means to acquire food. Sen devised methods of measuring poverty (Sen Index) that yielded useful information for improving economic conditions for the poor.

C.T. Kurien was born in 1932 and is a professor of economics, now retired and resident in Bangalore, who has written many books on economics. Although left, he is far from being a Marxist. C. T. Kurien attended Madras Christian College, graduating in 1953 with a Master's in Economics. He went on to Stanford University where he was awarded a Ph.D in 1962. Returning to Madras Christian College, between 1962 and 1978 he was Professor and Head of the Department of Economics. Kurien was a National Fellow of the University Grants Commission from 1975 to 77. In 1978 he was appointed Director of the Madras Institute of Development Studies, a national centre for social science research, holding this position for ten year. He was a National Fellow of the Indian Council of Social Science Research from 1992 to 1994. He was chairman of the Madras Institute of Development Studies from 1997 to 2003. He was President of the Indian Economic Association in 2002. As of 2012 he was a member of the board of the Institute for Social and Economic Change (ISEC) in Karnataka. His most recent book "WEALTH AND ILLFARE – An Expedition into Real Life Economics" was published in 2012, when the author was 80. It provides an accessible overview of economics, but questions the value of an economic system based on the pure pursuit of profit. In 1978 he took over the Directorship of the Madras Instituted of Development Studies founded by Malcolm Adiseshiah in 1970 and reconstituted in 1977 as a national centre for social science research within the framework of the Indian Council of Social Science Research. He was Director for a decade during which he built up the place as a noted institution combining rigour and relevance in its academic programme. He served as Chairman of the Institute from 1997 to 2003. He was Professor and Head of the Department of Economics in the Madras Christian College from 1962 to 1978. Dr. Kurien's writings - consisting of 15 highly commended books, many research papers and popular articles. His book 'Poverty, Planning and Social Transformation' is very popular.

Sukhamoy Chakravarty was an Indian economist who along with Prasanta Chandra Mahalanobis was the key architects of the Five-Year plans of India. After a brilliant undergraduate career, Professor Chakravarty went to Rotterdam to work for a Ph.D. under Jan Tinbergen. His dissertation was on the theory of planning. Professor Chakravarty was a mathematical economist; he combined that subject with political economy. He had a special affinity to the classical economists and the Cambridge School of Sraffa, Joan Robinson and Kaldor. He assumed a teaching post in Massachusetts Institute of Technology but returned to India to join the Planning Commission. He was also a professor at the Delhi School of Economics. Prof Chakravarty thought that the government should invest in agriculture, so as to spread the 'green revolution' technology from Punjab and Haryana to the rest of the country. He argued that this would lead to a large increase in agricultural productivity and to a rise in rural living standard. Towards the end of his life he was also deeply concerned with the more general theoretical, as well as practical policy questions facing not only India but all developing countries today. He passed away on 22nd August 1990. He is known for Chakravarty Committee regarding his recommendations for reforming banks.

Krishna Bharadwaj was born in Karwar, Karnataka in 1935. She was Indian economist who has contributed mainly in the field of economic theory. She is known for her understanding of the classical and neoclassical economics. She has revived the classical and new classical ideas in her writings. After studying postgraduate degree in Bombay she completed her Ph.D. in Transport Economics in 1960. Bharadwaj was the editor of the *Economic Weekly*. She has thorough understanding of the classical theory of value and distribution.Prabhat Patnaik is an Indian Marxist economist and political commentator. He taught at the Centre for Economic Studies and Planning in the School of Social Sciences at Jawaharlal Nehru University in New Delhi, from 1974 until his retirement in 2010. He was the Vice-Chairman of the Planning Board of the Indian state of Kerala from June 2006 to May 201. Prabhat Patnaik was born in Jatni in Orissa in September 1945 and after early schooling in his home town, studied at Daly College, Indore, on a Government of India Merit Scholarship. He passed his B.A. with Economics Honours from St. Stephen's College, Delhi. He went to Oxford University in 1966 on a Rhodes Scholarship and studied at Balliol College and later at Nuffield College. He obtained his B.Phil and his D.Phil. degrees from Oxford University. Patnaik joined the Faculty of Economics and Politics of the University of Cambridge, UK in 1969 and was elected a Fellow of Clare College, Cambridge. In 1974 he returned to India as an Associate Professor at the newly established Centre for Economic Studies and Planning (CESP) at the Jawaharlal Nehru University (JNU), New Delhi. He became a Professor at the Centre in 1983 and taught there till his retirement in 2010. His specialization is Macroeconomics and Political Economy, areas in which he has written a number of books and articles. He served as the Vice-Chairman of the Kerala State Planning Board from June 2006 to May 2011. Prabhat Patnaik is astaunch critic of neoliberal economic policies, and is known as a social scientist of Marxist- Leninist persuasion. According to him, the higher the economic growth the more will be the increase in the magnitude of absolute poverty. The only solution is to alter the class orientation of the State.

Jagdish Natwarlal Bhagwati was born in 1934, into a Gujarati family in the Bombay Presidency during the British Raj, and graduated from Sydenham College, Mumbai. He then went with "senior status" to read over two years for the BA in Economics at Cambridge. Jagdish Natwarlal Bhagwati is an economist and professor of economics and law at Columbia University. Bhagwati is a Democrat. He is well known for his research in international trade and for his advocacy of free trade. He received the Ph.D. in Economics from the Massachusetts Institute of Technology in 1967. Bhagwati has previously served as an external advisor to the Director General of the World Trade Organization in 2001, as a special policy advisor on globalization to the United Nations in 2000, and as an economics policy advisor to the Director-General of the Trade from 1991 to 1993. From 1968 until 1980, Bhagwati was an economics professor at the Massachusetts. Bhagwati currently serves on the Academic Advisory Board of Human Rights Watch (Asia) and on the board of scholars of the Centre for Civil Society. He is a Senior Fellow of the Council on Foreign Relations. In January 2004, Bhagwati published *In Defense of Globalization,* a book in which he argues "this process (of globalization) has a human face, but we need to make that face more agreeable." In 2006, Bhagwati was a member of the Panel of Eminent Persons who reviewed the work of the United Nations Conference on Trade and Development (UNCTAD).

Prof. Amit Bhaduri after his degrees in Economics from the Universities of Calcutta and Cambridge received Ph.D. from Cambridge University. He was awarded Stevenson Prize for the best piece of research in that University in 1966. He was extended Honorary Life Fellowship of the Centre for Development Studies, Trivandrum, in 1974. He was given the ICSSR Prize in recognition of original work in Economics in 1983. He was nominated to Honorary Life Professorship of Jawaharlal Nehru University in 1986. He held teaching positions in the Presidency College, Kolkata; Pembroke College, Cambridge; Delhi School of Economics, University of Delhi; Centre for Development Studies, Trivandrum; Stanford University, U.S.A. and many other Indian and foreign universities and institutions. Prof. Bhaduri acted as Research Consultant to ILO, UNCTAD, UNDP, UNIDO, U.N., and Economic Commission for Asia and the Pacific, Economic Commission for the Middle East; Vienna Institute of Comparative Economic Studies; and Development Bank of South Africa. Prof. Bhaduri has 5 books to his credit and has published more than 70 research articles in international journals of repute.

14
Dialectical Materialism

14.1. Dialectical Materialism

Marx's theory of social development is also known as 'dialectical materialism.' Marxian 'materialism implies that ideas, institutions, philosophies, religions, etc., all are determined by real material condition of life. This materialism is also at the root of Marxian concept of social classes. Marxian materialism simply means that men's economic activities are fundamental and they determine the general way the character of everything else they do. Marx got the idea of 'dialectics' from the German philosopher Hegel. According to the dialectical principle everything includes it's opposite or negative. A thesis has an antithesis. The conflict of these opposites leads to the transformation of this thesis into a synthesis. In other words, the basic cause of change in all things in their self-movement, lies in the struggle of opposites or in their internal contradiction. The contradiction gives rise to a synthesis between the thesis and antithesis. Human societies proceed from stage to stage of their development so that within each stage fundamental contradictions arise. When these contradictions are resolved, the society passes from one stage to another.

Marx's future-oriented perspective has its basis in his materialist conception of history. He suggests that the ways societies provide for their material well-being affects the type of relations that people will have with one another, their social institutions, and the prevailing ideas of the day. Marx uses the term 'the forces of production' to refer to the ways in which people provide for their needs. He uses the term 'relations of production' to describe social relationships that dominate the productive capacities of a society. Under capitalism, the forces of production lead to a set of relations of production which pit the capitalist and the proletariat against one another. To change the relations of production, Marx felt revolution was necessary. Revolution arises from exploited classes agitating for change in the relations of production that favour transformations in the forces of production. Marx applied the doctrine of dialectical materialism to analyse the historical evolution of society. Transition from one stage of social development to another is not caused by new principles or new ideas of truth and justice. What actually happens is that, in course of time, as productive forces change, they come into conflict with the existing relations of production. Inherent contradictions develop. The relations of production instead of furthering progress rather hinder progress. A social revolution takes place. Owing to a change in the economic system of society the superstructure is thrown aside. The transition thus takes place through a struggle between the classes. The victor in each becomes the vanquished in the next. Marx has broadly distinguished four different stages through which the society has passed.

These are primitive communism, slavery, feudalism and capitalism. The stage of primitive communism is characterized by a classless society. The means of production are socially owned. If there was any division of labour, it was limited to family. In course of time pastoral habits developed. People learnt how to tame animals. People learnt weaving and rudimentary forms of mining developed. All these led to the development of exchange. With the change in the forces of production, there was a great increase in the demand for labour, to carry on the new productive activities. This could not be satisfied under the existing form of economic society. The widening field of production brought slavery into existence. In other words, primitive communism was changed into slavery. Under slavery a new relation of production is established. The slave owner owned the means of production. There was further division of labour. Exchange flourished. New Institutions in the form of private ownership of land, money, interest, etc., developed. The society became more complex. States emerged to bring about coordination among the divergent interests. In course of time productive forces further developed. Then slavery gave way to feudalism. As we have noted elsewhere under feudalism, the feudal lord partially owned the means of production. The serfs or tenants were tied to the land, but they also owned some property. In the countryside agriculture flourished. In the urban areas small manufacturing handicraft used to be privately owned and operated by craft men. Production and exchange expanded. The scale of production under feudalism was very small and limited, which camein conflict with the growing productive powers which were slowly emerging. Besides the feudal, social and political forms also proved to be a hindrance to the further growth of the production powers of the economy. For instance, in England, since the 16th Century the bourgeoisie gradually became stronger on the basis of ne productive forces, such as new machines, factory processes, technical knowledge, superior weaponry, etc. These productive forces enabled them to challenge the supremacy of older feudal classes whose power and position depended on land rights, control of rural work forces and the prevailing social and legal systems. Conflicts followed. There was a series of revolutions. Ultimately, the economic structure changed from feudalism to capitalism. There was a transformation of the class structure from one consisting of lords and serfs to one of capitalists and workers.

Under the capitalist mode of production, the capitalist owns the necessary means of production. The worker owns none. The worker owns none. He is forced to work for the capitalist. From small-scale production there is a change to large-scale production. Production is for the market to earn more and more profit. Division of labour becomes more and more complex. There is free competition and freed of enterprise. Capitalism also is not free from this process of change. Growth of industrial capitalism introduced machines on a vast scale. This led to large-scale production, centralization, monopoly, finance capital and colonialism. In due process this system also developed its internal contradictions. Side by side, the capitalist mode of production will create a class of exploited men and then, unite, and strengthen this proletariat class which will overthrow capitalism. Under the fifth stage of socialism, which has not arrived yet, the workers themselves will own the means of production. The inherent contradictions within capitalism will be eliminated. Production will reach its fullest development. In this process of historical development of mankind, each successive stage represents an advance upon the preceding stage. Once a system outlives its utility and becomes an obstacle to the further development of the, productive forces, it will disintegrate, in the midst of a social revolution. This social revolution does not take place due to any religious, political or ideological reason as people used to believe before Marx. Marx held that the fundamental cause that governed every great movement in history was an economic cause. And historical evolution took place in accordance with the dialectic principle.

14.2. Marxian Theory of Disintegration of Capitalism

Marx was of the view that capitalism was destined to face disintegration ultimately. Marx believed that capitalism contained the seeds of its own destruction. He held that capitalism will disintegrate because of its inherent conflicts and contradictions. Following are the inherent conflicts and contradictions:

a. Alienation (Marx's moral critique of capitalism)

Under capitalism, the relationship between labour and human expression changes: rather than labouring to fulfil their needs or express ideas, workers do so at the demands of capitalists. Workers are alienated from their labour because it no longer belongs to the worker, but rather to the capitalist. This alienates workers in four ways:

I. Workers are alienated from their productive activity, in that they no longer labour to satisfy their own needs.

II. Workers are alienated from the product of their labour, which now belongs to the capitalist. Instead of finding expression in producing, workers turn to consuming to express themselves.

III. The cooperative nature of work is destroyed through the organization of the labour process, alienating workers from their fellow workers. Additionally, workers often must compete against one another for work and pay.

IV. Workers are alienated from their human potential, as the transformative potential of labour is lost under capitalism.

b. Exploitation

Exploitation is a set of social relations on which capitalism is built. Capitalists exploit workers by paying them less in wages than the value they produce. While a worker may earn eight dollars a day in wages, s/he may produce ten dollars a day worth of value, creating what Marx called surplus value. Capital grows by exploiting workers to generate ever greater amounts of surplus value, usually by lowering workers' wages. In addition, capitalists constantly compete with one another over capital by finding new ways to generate profit and surplus value in order to maintain an edge. Marx calls this drive the general law of capitalist accumulation. Capitalism is not the only historical epoch in which individuals are exploited, but it is the only one in which the mechanisms of exploitation are hidden behind independent, objectified, and reified structures, such as the market.

c. Creation of Surplus Value

d. Falling Tendency of Rate of Profit

The main objective of the capitalists is to make profit. But in a capitalist society as capital accumulates, more proportion of it will consist of constant capital. Constant capital is the product of past labour and it does not create any new value. Only variable capital can create value. Although constant capital does not produce any surplus value yet the use of constant capital like machinery etc. increases the productivity of labour. So by using constant capital, the capitalist can reduce his cost, undersell his competitors and gain a temporary advantage over them. Competition therefore compels the capitalists' continually to increase their constant capital. Now, as the rate of profit is the ratio between surplus value and constant plus variable capital, the average rate of profit inevitably declines. Thus, in the short run it may be possible to offset the tendency of the rate of profit to fall by increasing the productivity of' labour. But in the long run the increasing organic composition of capital will bring about a sure decline in the rate of profit. The rate of profit is an indicator of the economic health of market capitalism. A declining rate of profit indicates that capitalism is in the process of decomposition.

e. Reserve Army of Labour and Unemployment

Marx assumed that under capitalism there was a tendency to produce a surplus labour force. He called it as the 'reserve army of the unemployment.' In the normal process of capitalist accumulation and production, in a mature capitalist economy, there is a tendency for a progressive decline in the amount of variable capital. This will lead to a decline in employment opportunities. With the growing use of machinery the demand for labour in the industries would progressively fall. Machines would increasingly devour men. The larger the quantity of capital per worker, the greater the productivity of labour, the smaller the number of workers employed to produce a certain quantity of output. With the increase in the size of individual enterprises consequent upon concentration and centralisation of capital, the proportion of constant to variable capital would increase. The recurring economic crises also will vastly add to the army of unemployed. All these factors would tend to create an industrial reserve army or relative surplus population under capitalism.

f. The Increasing Misery of the Proletariat

Marx further held that with the progress of capitalism there was a tendency towards pauperisation or immiserisation of labour. His doctrine of increasing misery of labour has been interpreted in various ways. First, the condition of the -proletariat worsened due to unemployment. Due to the presence of a vast 'reserve army' the wages of the workers would be maintained at the minimum level. Intense competition among the workers would reduce the price of labour. Competition would also lead to over-work which contributed to their high mortality rate. Thus sometimes there is an absolute material deterioration. The economic success of capitalism followed by high production and wealth intensifies proletarian misery. The capitalist tries to realise the maximum possible surplus value from the worker. As the rate of profit of the capitalist falls, he becomes ever more desperate to raise the rate of exploitation. He lengthens the working day to the maximum extent possible. It even 'usurps the time for growth, development and healthy maintenance of the body' of the worker. The worker's wife children are thrown 'beneath the wheels of the juggernaut of capital.' There is mental degradation, anguish and growth of insecurity. Thus the downtrodden workers will grow increasingly miserable from the economic point of view.

g. The Increasing Severity of Economic Crises

The internal contradictions of the capitalist system manifest themselves in periodic economic crises. As Capitalism grows, economic crises take place with increasing severity. However Marx did not develop any separate business cycle theory in a systematic manner. However from his overall business cycle analysis, we may deduce three distinct theories of capitalist crisis. These are, (i) crisis associated with the falling rate of profit, (ii) crisis arising from disproportionality and (iii) under-consumption theory of crisis. First, in the process of capitalist development there in a relative decrease of variable as compared to constant capital and total capital. There is a rising organic consumption of capital. The degree of labour exploitation remaining the same, the rate of surplus value expresses itself in a continuously falling average rate of profit. Under capitalism production is guided by profit motive. With falling tendency in the rate of profit further accumulation is affected. It also poses as a threat to the production process. It also promotes overproduction, speculation, crisis and surplus capital along with surplus population. Secondly, according to Marx, crises under capitalism occur from disproportionality. This disproportionality takes place due to two basic characteristics of the capitalist system. First, there is 'anarchy of production'. It means that m a capitalistic economy productive decisions are made by. A large number of independent producers. Each producer has to anticipate the market demand for himself. He has no knowledge about the moves of his rivals. The only factor to guide him is the profit motive. Secondly, there is no systematic plan or orderliness in capitalist production. There is absolutely no understanding or coordination among the various producers. Each stands for himself. The inevitable consequence is the bewildering lack of coordination among the different branches of production. Consequently, periodic disequilibrium arises between demand and supply. Economic crisis is thus inevitable in a capitalist economy and nothing in particular can be done to stop it. If we sincerely try to remedy the situation, capitalism would not be capitalism. To abolish crisis, capitalism must be abolished.

Thirdly, Marx also held that crises were caused by under consumption which had its roots in the inner contradictions of capitalism. The capitalist economy generates enormous productive powers. In fact, it has a tendency to develop the productive power on an ever-expanding scale. With this there is large-scale production of consumption goods, intended for sale. But under the prevailing pattern of distribution of income, the consuming power of the society in limited. Here by 'consuming power' we mean effective demand which depends on income. They receive subsistence wage. Unemployment further aggravates their purchasing power. The capitalistic distribution of income is in conflict with capitalistic production. The consumption power of the society is further restricted by the tendency of the capitalists to accumulate, so the demand for consumption goods cannot keep pace with the supply. In other words, there is an increasing gap between aggregate supply on the one hand and consumption demand on the other. Economic crises naturally follow.

h. Concentration and Centralisation of Capital

Another prediction of Marx regarding the future of capitalism is the concentration and centralisation of capital. Marx's theory of concentration speaks of the tendency of the capitalist process 'to increase the size both of industrial plants and of units of control.' The concentration of capital implies the tendency for capital as a whole to accumulate. The capitalist has an endless passion for accumulation. The capitalists converted a large part of their surplus value into capital, In turn more capital meant the ability to enjoy more workers, more exploitation, more surplus value, more capital and so on and on. Capital thus tended to increase in total size like a snowball rolling downhil1. According to Marx, under capitalism this process of capital accumulation would continue far into the future. Another development which would take place simultaneously is the centralisation of capital. It refers to the tendency of existing capital to be concentrated in the hands of fewer and fewer number of larger and larger capitalists. When capital increases the number of capitalists falls. This centralization takes place due to several reasons. There is the tendency for the rate of profit to fall. During the business cycles there is overproduction and huge stocks of goods remain unsold. With the increasing use of machinery and increase in labour productivity, goods become cheaper and cheaper. All these factors give rise to a severe competitive struggle among the producers. Competitive struggle serves as an agent of centralisation. It always ends in the ruin of many small capitalists, whose capitals pass into the hands of their conquerors, partly vanish. The strong either crushes or absorbs the weak one. One capitalist kills many. The expropriators are expropriated. The small capitalists who lose in the battle of competition, drop out of the race and join the ranks of the proletariat, with the elimination of the smaller enterprises, competition disappears. The economic scene comes to be dominated either by a single capitalist or a single giant corporation. Thecredit system further aggravates the process as finally it transforms itself into an immense social mechanism for the centralisation of capitals. In the development of capitalist production, concentration and centralisation become self-perpetuating.

i. Proletarianization of the Society

Capitalism reduces all intermediate classes to the level of the proletariat. According to Marx this would happen as ultimately the proletariat would swallow up all other classes; except the capitalists. Competition would grow more and more severe which would kill many small capitalists. Capitalism would invade every branch of productive activity. This would proletarianize the intermediate classes of artisans, shopkeepers and peasants.

j. Finance Capital and Capitalistic Imperialism

Concentration and centralisation of capita will lead to a stage of 'Finance capital'. In this stage of capitalist development monopolies will play a predominant role m the economic life, after replacing free competition. There is a merger of bank capital with industrial capital. There is an unholy alliance between monopolist producers and powerful bankers and financial interests. A financial oligarchy is created on the basis of this 'finance capital'. Production is controlled through credit. The monopolist now not only earns profits but super-profits. The capitalist reaches the peak of his parasitic existence. Finally, once finance capital becomes dominant domestically, the financial magnates reach out beyond the boundaries of individual capitalist countries. This stage is known as capitalistic imperialism. This is a stage of international capitalism. At this stage the contradictions created by capitalist production are not limited to' a single country. They assume an international character. Capitalistic imperialism bas been called the highest stage of capitalism by Lenin, because all the contradictions inherent in the system of capitalist production assume their full and ultimate form with it.

k. The Eventual Breakdown and Inevitability of Socialism

According to Marx, the process of accumulation under capitalism inevitably leads to certain contradictions. There is an alienation and exploitation of labour. There is a secular tendency for the rate of profit to decline. Business cycles or economic crises occur regularly and with increasing ferocity. Economic power is concentrated in the hands of fewer and fewer capitalists. The economic life comes to be dominated by giant monopolistic and oligopolistic organisations. There is a steady deterioration in the condition of the workers. Thus inherent contradictions of the capitalist system come to the surface rendering its operation increasingly difficult. Life gradually becomes intolerable. At last there would be so serious a catastrophe that the capitalist system would destroy itself. This will mark the beginning of a revolutionary period during which the working class, at once oppressed and disciplined by its special position in society, will overturn the existing relations of production and establish in their stead higher Socialist relations of production. This is how through a socialist revolution the capitalist order would be inevitably transformed into socialism. Thus Marx's economic analysis flows into the sociology of class and socialism. Thus, we find that capitalism will not finally collapse by itself in an automatic mechanical way. The destruction of capitalism must be the work of human agency. The key to social transformation lies in the antagonistic relations between classes. This transformation is brought about by the proletariat with a conscious effort-through violence and revolution.

3. Criticisms of Marx's Theory

Following are criticisms that are commonly made:

- Too much emphasis is given to the economic factor in explaining social order and change. Culture seemed to be explained solely as derived from the economic 'substructure'. However it has a degree of 'autonomy'; for example it is difficult to explain the advent of gay liberation in terms of productive or economic relations.
- Even if you get rid of capitalism you might still have enormous problems of conflict and domination in society. State bureaucracies as well as capitalists can dominate – ask the Russians.
- Marx's theory of history is contradicted by the fact that industrialised countries have not moved closer to revolution. The recent revolutions have been in peasant societies, such as China. Capitalist societies seem to have become more secure from threat of revolution throughout the 20th century.
- Many would say there are no laws of history and that Marx was mistaken in thinking he had discovered the laws of history, and in thinking that his theory was scientific.
- Anarchists say Marxists fail to grasp the unacceptable dangers in their readiness to take an authoritarian-centralist approach. Marxists are willing to use the authoritarian state to run society after the revolution and to be ruthless in this. This is extremely dangerous; those in control can't be trusted and are very likely to become an entrenched dictatorship. (*e.g.*, Stalinism.)
- Many if not all Anarchists would also reject Marx's theory of how capitalism can or will be replaced, which involves confronting capitalism, class conflict, seizing the state and taking power from the capitalist class, and destroying capitalism, a process which will probably involve violence. However some anarchists believe the change could come via increasing awareness and disenchantment, the building of alternative communities based on anti-capitalist principles, and thus an increase in the numbers who want to abandon capitalism especially given that its coming difficulties will probably increasingly reveal its inability to provide for all.
- Marx (and most Marxists today) failed to take ecological sustainability into account. They are strong believers in industrial development and 'progress', rising material 'living standards' and economic growth. They think that capitalism is responsible for all problems and that when it has been eliminated we can release the previously restricted power of industry and eliminate waste to enrich everyone. In other words, Marxism has no concept of 'limits to growth'. Affluence and economic growth are regarded as desirable and possible. 'Dark green' critics insist that a good, post-capitalist society cannot be a growth society, and it cannot have high per capita levels of resource consumption. Getting rid of capitalism

is not enough; there is even bigger problem, set by the commitment to industrialism, growth and affluence. Marx could not have known that a time would come when we would run into a problem of over-consumption. From the perspective of 'The Simpler Way' a high quality of life for all is achievable without high material 'living standards' or much modern technology, let alone industrialisation and IT etc. We have no doubt that Marx was mistaken about scarcity, the stinginess of nature and the impossibility of human emancipation before technical advance delivered material abundance. His concept of development was really the same as capitalist modernisation. That is capital is crucial for development. He dismissed peasant ways and Marxists are not sympathetic to the notion of 'appropriate development' conceived mainly in terms of 'subsistence' and low/intermediate technology and cooperative ways focused on local economic self-sufficiency a Gandhian way. In other words, advocates of The Simpler Way claim Marx was quite mistaken in thinking that socialism would not be possible without modern technology, industrialisation and material affluence. Achieving a good society does not require elaborate technology nor abundance. It depends on whether or not the right values are held. There have been societies, and there are societies today in which people live well with very humble material lifestyles and without modern technology.

☆ Marxist ideas on how to change society are also strongly criticised by the Anarchists. Marxists thought capitalism must be fought and overthrown through violent revolution, because the capitalist classes will never voluntary give up any of its privileges. There must be leadership by a vanguard party prepared to be ruthless and to use violence, and which will rule in an authoritarian way after the revolution. Eventually when a person have developed the right ideas and values the state can dissolve and there will be a communist society.

14.2. List's Theory of National Economics

List's theory of "national economics" differed from the doctrines of "individual economics" and "cosmopolitan economics" by Adam Smith and J.B. Say. List contrasted the economic behaviour of an individual with that of a nation. An individual promotes only his own personal interests but a state fosters the welfare of all its citizens. An individual may prosper from activities which harm the interests of a nation. "Slavery may be a public calamity for a country, nevertheless some people may do very well in carrying on the slave trade and in holding slaves." Likewise, activities beneficial to society may injure the interests of certain individuals. "Canals and railroads may do great good to a nation, but all waggoners will complain of this improvement. Every new invention has some inconvenience for a number of individuals, and is nevertheless a public blessing". List argued that although some government action was essential to stimulate the economy, an overzealous government might do more harm than good. "It is bad policy to regulate everything and to promote everything by employing social powers, where things may better regulate themselves and can be better promoted by private exertions; but it is no less bad policy to let those things alone which can only be promoted by interfering social power."

Due to the "universal union" that nations have with their populace, List stated that "from this political union originates their commercial union, and it is in consequence of the perpetual peace thus maintained that commercial union has become so beneficial to them. The result of a general free trade would not be a universal republic, but, on the contrary, a universal subjection of the less advanced nations to the predominant manufacturing, commercial and naval power, is a conclusion for which the reasons are very strong. A universal republic., *i.e.* a union of the nations of the earth whereby they recognise the same conditions of right among themselves and renounce self-redress, can only be realised if a large number of nationalities attain to as nearly the same degree as possible of industry and civilisation, political cultivation and power. Only with the gradual formation of this union can free trade be developed, only as a result of this union can it confer on all nations the same great advantages which are now experienced by those provinces and states which are politically united. The system of protection, inasmuch as it forms the only means of placing those nations which are far behind in civilisation on equal terms with the one predominating nation", appears to be the most efficient means of furthering the final union of nations, and hence also of promoting true freedom of trade. In his seventh letter List repeated his assertion that economists should realise that since the human race is divided into independent states, "a nation would act unwisely to endeavour to promote the welfare of the whole human race at the expense of its particular strength, welfare, and independence. It is a dictate of the law of self-preservation to make its particular advancement in power and strength the first principles of its policy". A country should not count the cost of defending the overseas trade of its merchants. And "the manufacturing and agricultural interest must be promoted and protected even by sacrifices of the majority of the individuals; if it can be proved that the nation would never acquire the necessary perfection. without such protective measures.

List argued that statesmen had two responsibilities: "one to contemporary society and one to future generations". Normally, most of leaders' attention is occupied by urgent matters, leaving little time to consider future problems. But when a country had reached a turning point in its development, its leaders were morally obliged to deal with issues that would affect the next generation. "On the threshold of a new phase in the development of their country, statesmen should be prepared to take the long view, despite the need to deal also with matters of immediate urgency. List's fundamental doctrine was that a nation's true wealth is the full and many-sided development of its productive power, rather than its current exchange values. For example, its economic education should be more important than immediate production of value, and it might be right that one generation should sacrifice its gain and enjoyment to secure the strength and skill of the future. Under normal conditions, an economically mature nation should also develop agriculture, manufacture and commerce. But the two latter factors were more important because they better influenced the nation's culture and independence. These factors were especially connected to navigation, railways and high technology, while a purely agricultural state tended to stagnate. But, List claims, only countries in temperate regions were adapted to grow higher forms of industry. On the other hand, tropical regions had a natural monopoly in the production of certain raw materials. Thus, there was a spontaneous division of labour and confederation of powers between these two groups of countries. List contended that Smith's economic system is not an industrial system but a mercantile system, and called it "the exchange-value system". Contrary to Smith, he argued that the immediate private interest of individuals would not lead to the highest good of society. The nation stood between the individual and humanity, and was defined by its language, manners, historical development, culture and constitution. This unity must be the first condition of the security, well-being, progress and civilization of the individual. Private economic interests, like all others, must be subordinated to the maintenance, completion and strengthening of the nation.

(a) Stages of Economic Development

List the raised that nations of the temperate zone (which are furnished with all the necessary conditions) naturally pass through stages of economic development in advancing to their normal economic state. These are:

- ✰ Pastoral life
- ✰ Agriculture
- ✰ Agriculture united with manufactures
- ✰ Agriculture, manufactures and commerce are combined

The progress of the nation through these stages is the task of the state, which must create the required conditions for the progress by using legislation and administrative action. This view leads to List's scheme of industrial politics. Every nation should begin with free trade, stimulating and improving its agriculture by trade with richer and more cultivated nations, importing foreign manufactures and exporting raw products. When it is economically so far advanced that it can manufacture for itself, then protection should be used to allow the home industries to develop, and save them from being overpowered by the competition of stronger foreign industries in the home market. When the national industries have grown strong enough that this competition is not a threat, then the highest stage of progress has been reached; free trade should again become the rule, and the nation be thus thoroughly incorporated with the universal industrial union. What a nation loses in exchange during the protective period, it more than gains in the long run in productive power. The temporary expenditure is analogous to the cost of the industrial education of the individual.

"In a thousand cases the power of the State is compelled to impose restrictions on private industry. It prevents the ship owner from taking on board slaves on the west coast of Africa, and taking them over to America. It imposes regulations as to the building of steamers and the rules of navigation at sea, in order that passengers and sailors may not be sacrificed to the avarice and caprice of the captains. Everywhere does the State consider it to be its duty to guard the public against danger and loss, as in the sale of the necessaries of life, so also in the sale of medicines. While List once had urged Germany to join other 'manufacturing nations of the second rank' to check Britain's 'insular supremacy', by 1841 he considered that the United States and Russia would become the most powerful countries-a view expressed by Alexis de Tocqueville the previous year. List hoped to persuade political leaders in England to co-operate with Germany to ward off this danger. His proposal was perhaps not so far fetched as might appear at first sight. In 1844, the writer of an article in a leading review had declared that 'in every point of view, whether politically or commercially, we can have no better alliance than that of the German nation, spreading as it does, its 42 millions of souls without interruption over the surface of central Europe.

The practical conclusion which List drew for Germany was that it needed for its economic progress an extended and conveniently bounded territory reaching to the seacoast both on north and south, and a vigorous expansion of manufacture and trade, and that the way to the latter lay through judicious protective legislation with a customs union comprising all German lands, and a German marine with a Navigation Act. The national German spirit, striving after independence and power through union, and the national industry, awaking from its lethargy and eager to recover lost ground, were favourable to the success of List's book, and it produced a great sensation. He ably represented the tendencies and demands of his time in his own country; his work had the effect of fixing the attention, not merely of the speculative and official classes, but of practical men generally, on questions of political economy; and his ideas were undoubtedly the economic foundation of modern Germany as applied by the practical genius of Bismarck List considered that Napoleon's 'Continental System', aimed just at damaging Britain during a bitter long-term war, had in fact been quite good for German industry. This was the direct opposite of what was believed by the followers of Adam Smith. As List put it: I perceived that the popular theory took no account of nations, but simply of the entire human race on the one hand, or of the single individual on the other. I saw clearly that free competition between two nations which are highly civilised can only be mutually beneficial in case both of them are in a nearly equal position of industrial development, and that any nation which owing to misfortunes is behind others in industry, commerce, and navigation. must first of all strengthen her own individual powers, in order to fit herself to enter into free competition with more advanced nations. In a word, I perceived the distinction between cosmopolitical and political economy List's argument was that Germany should follow actual English practice rather than the abstractions of Smith's doctrines: Had the English left everything to itself–'Laissez faire, laissez aller', as the popular economical school recommends–the [German] merchants of the Steelyard would be still carrying on their trade in London, the Belgians would be still manufacturing cloth for the English, England would have still continued to be the sheep-farm of the Hansards, just as Portugal became the vineyard of England, and has remained so till our days, owing to the stratagem of a cunning diplomatist. Indeed, it is more than probable that without her commercial policy England would never have attained to such a large measure of municipal and individual freedom as she now possesses, for such freedom is the daughter of industry and wealth.

15
Thornstein Veblen

15.1. Thornstein Veblen

Thornstein Veblen was an American economist and sociologist, and leader of the institutional economics movement. Besides his technical work he was a popular and witty critic of capitalism, as shown by his best known book The Theory of the Leisure Class (1899). Veblen is famous in the history of economic thought for combining a Darwinian evolutionary perspective with his new institutionalist approach to economic analysis. He combined sociology with economics in his masterpiece The Theory of the Leisure Class (1899) where he argued that there was a fundamental split in society between those who make their way via exploitation and those who make their way via industry. In early barbarian society this is the difference between the hunter and the gatherer in the tribe, but as society matures it is the difference between the landed gentry and the indentured servant. The titular manifestation of those with the power of exploit is the "Leisure Class" which is defined by its lack of productive economic activity and its commitment to demonstrations of idleness. As Veblen describes it, as societies mature, conspicuous leisure gives way to "conspicuous consumption", but both are performed for the sole purpose of making an invidious distinction based on pecuniary strength, the demonstration of wealth being the basis for social status.

Veblen was sympathetic to state ownership of industry, but he had a low opinion of workers and the labour movement and there is disagreement about the extent to which his views are compatible with Marxism, socialism or anarchism. As a leading intellectual of the Progressive Era, he made sweeping attacks on production for profit, and his stress on the wasteful role of consumption for status greatly influenced socialist thinkers and engineers who sought a non-Marxist critique of capitalism. Fine (1994) writes that economists at the time complained that his ideas, while brilliantly presented, were crude, gross, fuzzy, and imprecise; others complained that he was a wacky eccentric. Scholars continue to debate what exactly he meant in his convoluted, ironic and satiric essays; he made heavy use of examples of primitive societies, but many examples were pure invention. He studied economics under John Bates Clark, a leading neoclassical economist, but rejected his ideas. He did his graduate work at Johns Hopkins University under Charles Sanders Peirce, the founder of the pragmatist school in philosophy, and at Yale University under laissez-faire proponent William Graham Sumner. Veblen is best known for his book *The Theory of the Leisure Class,* which introduced the term "conspicuous consumption" (referring to consumption undertaken to make a statement to others about one's class or accomplishments). This term, more than any other, is what Veblen is known for. Some types of luxury goods, such as high-end wines, designer handbags, and luxury cars, are considered Veblen goods, in that decreasing their prices decreases people's preference for buying them because they are no longer perceived as exclusive or high-status products. Similarly, a price increase may increase that high status and perception of exclusivity, thereby making the good even more preferable. Often such goods are no better or are even worse than their lower priced counterparts.

At the other end of the spectrum, where luxury items priced equal to non-luxury items of lower quality, all else being equal more people would buy the luxury items, even though a few Veblen-seekers would not. Thus, even a Veblen good is subject to the dictum that demand moves conversely to price, although the response of demand to price is not consistent at all points on the demand curve. Veblen developed a 20th-century evolutionary economics based upon Darwinian principles and new ideas emerging from anthropology, sociology, and psychology. Unlike the neoclassical economics that was emerging at the same time, Veblen described economic behavior as socially determined and saw economic organization as a process of ongoing evolution. Veblen strongly rejected any theory based on individual action or any theory highlighting any factor of an inner personal motivation. Such theories were according to him "unscientific." This evolution was driven by the human instincts of emulation, predation, workmanship, parental bent, and idle curiosity. Veblen wanted economists to grasp the effects of social and cultural change on economic changes. In The Theory of the Leisure Class, the instincts of emulation and predation play a major role. People, rich and poor alike, attempt to impress others and seek to gain advantage through what Veblen coined "conspicuous consumption" and the ability to engage in "conspicuous leisure." In this work Veblen argued that consumption is used as a way to gain and signal status. Through "conspicuous consumption" often came "conspicuous waste," which Veblen detested.

In The Theory of Business Enterprise, which was published in 1904 during the height of American concern with the growth of business combinations and trusts, Veblen employed his evolutionary analysis to explain these new forms. He saw them as a consequence of the growth of industrial processes in a context of small business firms that had evolved earlier to organize craft production. The new industrial processes impelled integration and provided lucrative opportunities for those who managed it. What resulted was, as Veblen saw it, a conflict between businessmen and engineers, with businessmen representing the older order and engineers as the innovators of new ways of doing things. In combination with the tendencies described in The Theory of the Leisure Class, this conflict resulted in waste and "predation" that served to enhance the social status of those who could benefit from predatory claims to goods and services.

Veblen generalized the conflict between businessmen and engineers by saying that human society would always involve conflict between existing norms with vested interests and new norms developed out of an innate human tendency to manipulate and learn about the physical world in which we exist. He also generalized his model to include his theory of instincts, processes of evolution as absorbed from Sumner, as enhanced by his own reading of evolutionary science, and Pragmatic philosophy first learned from Peirce. The instinct of idle curiosity led humans to manipulate nature in new ways and this led to changes in what he called the material means of life. Because, as per the Pragmatists, our ideas about the world are a human construct rather than mirrors of reality, changing ways of manipulating nature lead to changing constructs and to changing notions of truth and authority as well as patterns of behavior (institutions). Societies and economies evolve as a consequence, but do so via a process of conflict between vested interests and older forms and the new. Veblen never wrote with any confidence that the new ways were better ways, but he was sure in the last three decades of his life that the American economy could, in the absence of vested interests, have produced more for more people.

In the years just after World War I he looked to engineers to make the American economy more efficient. In addition to The Theory of the Leisure Class and The Theory of Business Enterprise, Veblen's monograph "Imperial Germany and the Industrial Revolution", and his many essays, including "Why is Economics Not an Evolutionary Science," and "The Place of Science in Modern Civilization," remain influential. Veblen did not reject economists' answers to the questions they posed; he simply thought their questions were too narrow. Veblen wanted economists to try to understand the social and cultural causes and effects of economic changes. What social and cultural causes were responsible for the shift from hunting and fishing to farming, for example, and what were the social and cultural effects of this shift? Veblen was singularly unsuccessful at getting economists to focus on such questions. His failure may explain the sarcastic tone his writing took toward his fellow economists. Veblen had to struggle to stay in academia. In the late nineteenth century many universities were affiliated in a substantial way with churches. Veblen's skepticism about religion and his rough manners and unkempt appearance made him unattractive to such institutions. As a result, from 1884 to 1891 Veblen lived on the largesse of his family and his wife's family. His big break came in 1892 when the newly formed University of Chicago hired his mentor, J. Laurence Laughlin, who brought Veblen with him as a teaching assistant.

Veblen later became the managing editor of the *Journal of Political Economy,* which was and is edited at the University of Chicago. Veblen spent fourteen years at Chicago and the next three at Stanford. He died in obscurity in 1929.

15.2 Production and Long-Period Competitive Supply

In deriving the long-period supply curve of a commodity, Marshall envisages production as organized by firms, typically family businesses. Each firm strives to minimize its production costs, substituting one productive factor or production method for another according to the "Principle of Substitution." In its simpler forms this involves marginalist adjustment to bring relative marginal value products into line with relative marginal costs. Marshall's firms do not have costless access to a common production function, but must grope and experiment their way to cost-reducing modifications. The long period supply curve is defined for a given state of *general* scientific and technical knowledge. The long-period supply price per unit of output at which such an industry can supply any quantity of output must just cover the cost of maintaining that level of output indefinitely. That is, it must just suffice to pay all the inputs (including management) needed to produce that level of output in a cost-minimizing way at rates which will just ensure that the requisite input quantities will continue to be forthcoming. In the case of skilled workers, in particular, the rate must just suffice to induce parents to apprentice new workers to the industry at a rate exactly offsetting the attrition through retirement, etc. Similarly, the return to fixed capital must just suffice to induce replacement of the existing stock of fixed assets, while the return to management must keep up the necessary replacement flow of managers. On the other hand, the return to land services must just suffice to prevent these services from migrating elsewhere, replacement not being necessary. As the level of industry output being considered is increased, the supply price will probably rise, mainly because of the need to pay a higher return to land so as to attract a greater supply from other uses, but perhaps also because of the need to pay more for rare natural talents which, like land, must be attracted in greater quantity from other uses, not being capable of replication through education and training. Such a tendency for long-period supply price to rise with output may be mitigated though seldom eliminated by substitution against inputs whose supply price is rising, and by possible external economies which increase each firm's efficiency by influences which depend, not on its own output, but on the entire industry's output. A tendency for supply price to rise with output will imply that inframarginal units of those inputs whose supply prices are rising receive rents, since all units will be remunerated at the rate necessary to induce continuing supply of the marginal unit.

(a) Representative Dirm

It is here that Marshall's "representative firm" enters the picture. It is best regarded as a parable which avoids the need to consider the entire distribution of firms. By definition, the long-period supply price of any level of industry output is the average cost of the representative firm at that level of output. Industry-level magnitudes may then be regarded as if they were generated by a fixed number of unchanging representative firms rather than by the actual heterogeneous body of ever-changing firms– that is, the manufacturing case may be treated as if it were an agricultural case. Such arguments add nothing conceptually and are prone to confuse, although it might be noted that Marshall believed an acute well-informed observer could select an actual firm which was close to being representative in this sense.

The average cost and size of the representative firm will change as *industry* output changes. There are two main reasons for this. A larger industry output is likely to generate more *external* economies, lowering the costs of every firm. But more importantly, the larger is industry demand the easier it will be for a new firm to build up a market, and so the larger the size to which firms will grow before they begin to decline. This will bring about greater access on average to unexhausted *internal* economies of scale, again leading to lower costs on average. For both these reasons, long-period supply price is likely to decline as a larger industry output is considered, even though the opportunity cost of obtaining greater supplies of land services and rare natural talents may rise. Again, the particular expenses curve may be used to display the surpluses or rents accruing to such scarce factors at any given level of industry output, but the relationship of this family of curves to the long-period supply curve is tenuous and complex. Rent obviously cannot be represented by a "triangle" above the supply curve when the latter is falling.

The conception of competition in Marshall's manufacturing case is much closer to later ideas of imperfect or monopolistic competition than to modern notions of perfect competition. Products are differentiated and firms are not pricing takers, but face at any time downward-sloping demand curves in their special markets. Even if the difficulties of rapidly building up a firm's internal organization can be overcome, the resulting enlarged output cannot be sold at a price covering cost–even granted substantial scale economies in production–without going through the slow process of building up a clientele and shifting the firm's particular demand curve. The time this takes is assumed to be considerable relative to the duration of the firm's initial vitality. But in some cases the difficulties of rapid expansion may be overcome. They may not have been very severe, as when different firms' products are highly substitutable, or the firm's founder may have unusual genius. In such cases the industry will pass into a monopoly or be dominated be a few, strategically-interacting firms, or "conditional monopolies" as Marshall termed them. Marshall's reconciliation of persisting competition with increasing returns and falling supply price is complex and problematic, but it does not depend in any essential way on scale economies being external to the firm. The concept of external economies is one of his significant contributions, although his treatment of it can hardly be called pellucid. But it was added more for verisimilitude than because it was theoretically essential to the structure of his theory.

(b) Price Determination and Period Analysis

The long-period supply curve for any good indicates for each market quantity the least price at which that quantity will continue indefinitely to be supplied. The equilibrium price and quantity (long period) are determined by the intersection of this supply curve with the negatively sloped market demand curve, indicating the highest uniform price at which any total quantity can be sold. In an agricultural case, equilibrium will be unique as the supply curve slopes positively. But in a manufacturing case, the supply curve, as well as the demand curve, will have negative slope, so that multiple equilibria can occur. Equilibrium is adjudged locally stable if demand price is above (below) supply price at a quantity just below (above) the equilibrium quantity. The intuitive justification for this is that the actual price of any available quantity is determined by the demand price, while quantity produced tends to increase (through both expansion of existing firms and rapid entry of new firms) whenever an excess of market price over supply price promises high profits, while it tends to decrease in the opposite case.

Marshall emphasized three broad cases. Temporary or market equilibrium analysis proceeded on the assumption of a fixed stock of output already available or in the pipeline. Short-period–normal equilibrium analysis permitted output to be varied but not the stock of productive "appliances" available to produce that output. "Appliances" must be taken here to cover skilled labour and business organization as well as fixed capital assets, so that the existing set of firms is to be taken as given. Finally, long-period–normal equilibrium, which has already been considered, allows the stock of appliances to be freely varied, as well as the level of output. In this case equilibrium incorporates the conditions necessary for inducing a replacement flow of each kind of appliance, including a replacement flow of new firms in the manufacturing case. Temporary equilibrium for a perishable commodity is simply a matter of selling off the existing stock. Marshall recognizes the possibility of "false trading"–sales at a non-equilibrium price–but argues that (a) this will not affect the eventual price if the marginal utility of money is constant, and (b) price will quickly settle close to that uniform price which would just clear the market if used in all transactions. With a storable good there is the further speculative possibility of holding back supply for future sale, and this gives expected future cost of production an indirect role in influencing current market price. Cost of production already incurred is an irrelevant bygone, however. In short-period–normal equilibrium, output is adapted to demand within the constraints set by the fixed supply of available "appliances." High demand will raise equilibrium output, but only within the limits possible by working existing appliances more intensively or pulling in versatile unspecialized labour and land from elsewhere. Low demand will lead to low utilization of appliances, perhaps idleness of some, and migration of unspecialized inputs to elsewhere. In the agricultural case a firm will change output until marginal prime or variable cost equals market price. In the manufacturing case, a fear of spoiling the future market or invoking retaliation from competitors tends to make a firm's output more responsive to variation in market price, and hence to make market price less responsive to demand shifts. Otherwise, the two cases are similar, both involving a fixed population of firms and a rising supply curve.

The return received by an appliance will often exceed the minimum necessary to induce its operation at the chosen intensity (its prime cost) and this excess is a "quasi rent." To the extent that land and rare natural talents are immobile in the short period, or less mobile in the short period than the long, then their returns too will often have a quasi-rent element. Otherwise, they will receive only differential rents, though often at rates differing from their long-period values. It should be stressed that the concepts of quasi rent and differential rent are relative to a specific use. The prime cost necessary to retain an input in this use may itself include rent or quasi rent when viewed in the context of a more inclusive set of alternative uses. Thus, from the viewpoint of all possible uses in the economy, the return to any factor in fixed supply is entirely a rent or quasi rent (the latter if fixity is only short-period). Marshall's period analysis, and more generally his partial-equilibrium approach to price determination, was designed in large part as a usable tool for the analysis of concrete issues. Its longevity amply testifies to its usefulness in this respect. But it was also meant to serve the more doctrinal purpose of clarifying the respective roles utility and cost of production play in determining value. The aim was to show that the greater the scope for supply adjustment permitted in the definition of equilibrium the more dominant the supply-side influence on price becomes. This doctrinal goal helps to account for the rather heavy weight given to long-period analysis in the *Principles*. For, as Marshall recognized, its value as a tool of applied analysis is seriously qualified by the fact that "violence is required for keeping broad forces in the pound of Caeteris Paribus during, say, a whole generation, on the ground that they have only an indirect bearing on the question in hand". That is, there is no good ground for assuming that background forces such as technology and tastes will remain constant for the length of time required for long-period equilibrium to be practically relevant. For concrete analysis of problems of such long duration it will often be necessary to transcend the period analysis, with its reliance on statical equilibrium, and undertake directly an analysis of secular change, of which Book VI, Ch. XII of the *Principles* on the "General Influence of economic progress" offers the main example, but not a very impressive one.

15.3. Welfare Economics

To serve as a tool of welfare economics, monetary measures of changes in consumer surplus, producer surplus and rent must be aggregated over individuals, but how are the resulting sums to be interpreted? Marshall's very limited and proximate attempts at formal welfare arguments are carried out within a utilitarian framework, for which the goal is maximizing aggregate utility. He implies that interpersonal utility comparisons are possible in principle and that utility functions will be similar for all members of any group that is homogeneous in terms of mental, physical and social attributes. Within such a group, the marginal utility of money will be the same for two individuals having the same income, and lower for the richer of two individuals having different incomes, assuming in each case that both individuals face the same trading opportunities. A postulated change (*e.g.* a government action) will impose gains and losses on individuals which can be measured and aggregated in money-equivalent terms, but how can these measures be translated into statements about aggregate gains and losses of utility? Marshall emphasizes two special cases. If the gains are distributed over groups, and over income classes within each group, in exactly the same proportions as the losses are distributed, then aggregate utility gain will stand in the same proportion to aggregate monetary gain as aggregate utility loss stands to aggregate monetary loss. Even without knowing this proportion, the aggregate net monetary gain or loss will serve as an index of the aggregate utility gain or loss (although it can only rank alternatives having the same *relative* distributions of monetary benefits and costs as the case in question). Alternatively, if money gains and losses are distributed similarly over groups, but within each group the gains accrue to individuals of lower income than those bearing the costs, then there must be an aggregate net utility gain even if the aggregate net monetary gain is zero-a warrant for certain redistributive policies. Marshall believed that these special cases were of quite wide applicability. In other cases, he saw that careful judgemental assessments of the marginal utility of money to the various injured and benefited groups would be necessary, assessments which could be used to transform monetary gains and losses into utility measures. But he gave little indication as to how these assessments might be obtained in practice.

Marshall's best known and most successful foray into formal welfare analysis was his proof that total welfare might be increased by using the proceeds of a tax on an "agricultural" industry to subsidize a "manufacturing" industry. All comparisons involved long-period equilibria and relied on the validity of aggregated money-equivalent measures of gains and losses. He demonstrated that the gain in consumer surplus in the expanded decreasing-cost manufacturing industry might exceed the combined loss in consumer surplus and producer rents in the contracted increasing-cost agricultural industry. No formal account was taken of the possible gain in producer rent in the manufacturing industry as this merely made the argument *a fortiori*. The crucial point in this argument, as Marshall recognized, is that producers are not harmed by "a fall in price which results from improvements in industrial organization" (1920, p. 472). It is immaterial whether the improved organization of the enlarged manufacturing industry is due to external economies or to internal economies resulting from an increase in the size of the representative firm. Contrary to much subsequent opinion, Marshall's tax-subsidy argument is not necessarily dependent upon external economies. Another significant, but overlooked, welfare analysis provided by Marshall was that of a monopolistic public enterprise in a situation where taxation involves an excess burden (1920, pp. 487– 93, 857–8). Marshall proposes the goal of "compromise benefit" which effectively sums consumer surplus and monopoly revenue, but the latter multiplied by the marginal cost of raising a unit of government revenue from other sources. Maximization of compromise benefit leads to the setting of what has come to be termed a Ramsey price. In the absence of an excess burden, when the marginal cost of extra government revenue is unity, this reduces to marginal cost pricing. The two examples of welfare analysis just described proceed within a partial equilibrium framework, treating each industry as negligible compared to the entire economy and regarding the marginal utility of money as approximately constant to each individual. The gains or losses to producers need only take account of the narrow differential advantages obtained by operating in the industry in question rather than in any other use. Marshall's rather fragmentary remarks on optimal tax systems, income redistribution, and the "doctrine of maximum satisfaction" cannot be restricted in this way, and so raise serious unresolved analytical difficulties. On the other hand, his tax-subsidy argument was offered as a valid counterexample to arguments that competition must lead to a social optimum, or that optimal indirect tax systems must involve uniform tax rates. It must also be borne in mind that utilitarian welfare economics was for Marshall only a proximate step towards a more evolutionary analysis of modes of improving the physical quality and the values and activities of mankind.

15.4. Inter-related Market and Distribution Theory

Marshall was anxious to emphasize the interdependence of markets and introduced his treatment of joint and composite demand and supply largely for this purpose. A group of products is jointly supplied or demanded if its members are all the outputs of or inputs into a single activity. A product is compositely supplied or demanded if it is produced or consumed by several activities. Marshall's formal treatment of joint demand and supply proceeded on the general assumption of fixed proportions, as did the related analysis of the derived demand for any one of several jointly-demanded inputs. The derived demand curve for an input is effectively constructed on the assumption that, as the price of the input in question is arbitrarily varied, prices of the remaining inputs and of the output of the activity in which they are used always adjust to keep demand equal to supply. The prime example of joint demand is the demand for productive inputs, and Marshall's analysis of market interdependence was carried through more fully in this specific connection, the role of substitution among inputs receiving full acknowledgement. The principle of substitution ensured that input usage was adjusted by firms to equate marginal value product, taken normally as the value of the marginal product (or "net product"), to the unit price of the input. Interdependence among input markets was further highlighted in the analysis of the competition of several industries for an input which is in temporarily or permanently fixed overall supply. A peculiarity of this last analysis was the insistence on excluding from the marginal cost of any industry the cost of bidding such fixed resources away from other uses. This is a perfectly legitimate application of the general envelope theorem: providing resource use is optimally adjusted, the marginal cost of increasing output will be the same whatever input or subgroup of inputs is increased. But Marshall's insistence on asymmetry where there is really symmetry can only be accounted for by his desire to legitimate, and extend to quasi rent, the classical doctrine that rent is price determined rather than price determining. Marshall's vision of market interdependence culminates in his treatment of distribution, where he seeks to bring out the extents to which the interests of different factors are consilient or conflicting. Distribution is determined by the interaction of the demands and supplies for the various inputs, the demands being essentially joint demands. Marginal productivity is a theory of input demand, not a complete theory of distribution, because the supplies of the various inputs cannot be viewed as fixed, at least in the long period. Indeed, in the long period the dominant influences on the prices of factors other than land are exerted by their supply conditions, the costs which have to be met in order for various kinds of labour and capital to just continue to be replaced in their existing uses and quantities. From an overall view "The net aggregate of all the commodities produced is itself the true source from which flow the demand prices for all these commodities, and therefore for the agents of production used in making them" (1920, p. 536). This aggregate, "the national dividend," is distributed among the factors of production. All have a common interest in increasing the size of the pie to be shared, but each has a selfish interest in restrictive practices which increase its own share even if they reduce the pie slightly. A prime question of social policy for Marshall is how these divergent incentives can be reconciled: how combined action by various groups, such as unions, can be prevented from assuming forms which, while perhaps individually beneficial to any one group in isolation, are certainly mutually harmful if undertaken by all.

Marshall here enters into macroeconomic forms of argument, and it is indeed true that he did toy with the formal specification of macroeconomic models of growth and distribution (Whitaker, 1975, Vol. II, pp. 305–16). But, with this exception, it should be emphasized that his treatment of market interdependence fell far short of a full theory of general equilibrium on Walrasian lines. Even when formalizing market interdependence in the mathematical appendix to the *Principles* (1920, pp. 846–56), he simply treated the demand or supply of each commodity as a function of nothing but the price of the commodity itself. The links between the generation of income in factor markets and the expenditure of that income in product markets were left quite vague. Again, it must be recalled that the development of comprehensive fully articulated equilibrium theories was not Marshall's aim.

Marshall's analysis of price and output determination by a profit-maximizing monopolist, and of the effects of taxing such a monopolist, followed the lead of Cournot. The concept of marginal revenue was implicitly in the mathematical statement, but Marshall's chosen vehicle was geometrical. Curves of average revenue and cost, and of their difference, average net revenue, y, (all functions of the quantity sold, x) were superimposed on a grid of iso-profit hyperbolae of form xy = constant. Profit was maximized when the average net revenue curve touched the highest such iso-profit curve. Weighting consumer surplus into the maximand, as well as net revenue, gave rise to the welfare analysis of "compromise benefit" already mentioned. Monopoly analysis was applied to trades unions, using the concept of the derived demand for an input, in which context Marshall laid down his four rules for inelastic derived demand. These were that the input should have no good substitutes, that the product it helps make should be inelastically demanded, that the input should account for only a small part of production costs, and that cooperating inputs should be inelastically supplied. A union controlling a labour input for which derived demand is inelastic can certainly raise wages–not only the wage rate but the total wages received–although at the price of unemployment of some members. Whether such a monopolistic restriction can be sustained for long is more doubtful, as there will be pressures both to enter the union and to evade its grasp by the relocation or reorganization of production. A more problematic question was whether "labour's disadvantage in bargaining" meant that combined action by workers could raise wages, even without any restriction of labour supply. Marshall believed that it did, but emphasized that the result might be less capital accumulation by non-workers, an outcome which could harm workers eventually. The extremes of monopoly and competition were both covered by the theory of normal value, even though the competition might be more akin to later concepts of imperfect or monopolistic competition than to any ideal form of perfect competition. But "normal action falls into the background, when Trusts are striving for the mastery of a large market" (1920, p. xiv). The incidents, tactics and alliances of oligopolistic conflict defied reduction to a simple general theory. They were to have been considered in the uncompleted second volume of the *Principles* and were to some extent covered by *Industry and Trade*. The latter's treatment of entry-limiting behaviour by a "conditional monopolist," who dominates the market but does not control entry, is of considerable interest in the light of much recent work on this class of problems.

15.5. Monetary Theory

Marshall was in full command of previous British discussions of monetary issues, but not himself a major contributor to the development of monetary theory. His evidence before Royal Commissions in 1887 and 1899 showed an impressive mastery of monetary analysis, both domestic and international, and was minutely examined by successive generations of Cambridge students, serving for many years virtually as a textbook. But it was not until 1923, with the appearance of *Money, Credit and Commerce*, that Marshall put forward his monetary views in a systematic way. By then these had not the novelty, nor the vigour, to advance contemporary discussion. Marshall's most important contribution to monetary theory was to place the overall demand for money in the context of individual choices as to the fraction of one's wealth to keep on hand as ready cash. This approach, set out clearly in a manuscript of the early 1870s, was developed by Marshall's Cambridge successors, especially A.C. Pigou and F. Lavington, into what is termed the "Cambridge k" approach. It laid the background for the treatment of the demand for money in J.M. Keynes. On international monetary theory, Marshall espoused a form of purchasing power parity.

Marshall's name is particularly associated with his proposals for "symmetalism," the use of a Fixed-weight combination of gold and silver as the monetary base, and for indexed contracts based on a "tabular standard of value," or price index, to be maintained by the government. The former was offered as an improvement on fixed-ratio bimetallism, of which he was never more than a lukewarm adherent. Marshall had interesting, if fragmentary, insights into business fluctuations and general unemployment, which he viewed as temporary disequilibrium consequences of credit market dislocations. These spilled over into general coordination failures, with unemployment in one market spreading to others by reducing demand in cumulative fashion–the germ at least of the multiplier concept. On the other hand, Say's law was maintained as an equilibrium truth of great importance. He saw the remedies for cyclical unemployment in the "continuous adjustment of means to ends, in such a way that credit can be based on the solid foundation of fairly accurate forecasts," and in curbs on reckless inflations of credit which are "the chief cause of all economic malaise".

16
International Trade

16.1. International Trade

Marshall's major contribution to international trade theory was his well-known geometrical analysis of the equilibrium and stability of two-country trade by means of intersecting offer curves. Each country's offer curve indicated the number of "bales" of home goods it was prepared to exchange for a specified number of bales of foreign goods, demand being elastic or inelastic as an increase in the latter caused the former to increase or decrease. Possibilities of multiple and unstable offer-curve intersections were noted. The offer curves themselves were taken as data, although complex readjustments of production and consumption underlay them. The need for a separate theory of international trade was justified, in classical vein, by the supposed international immobility of factors which remain mobile domestically. The main purpose of this theoretical apparatus was to examine the effects of tariffs. A country might gain by selfishly exploiting its monopoly power through restricting trade, and would certainly gain if trading equilibrium was on an inelastic portion of the foreign offer curve. But Marshall came increasingly to doubt the transferability of this result to a multi-country case, although admitting that it might apply to an export tax on an exceptional commodity (like British steam coal) lacking close substitutes and incapable of being produced elsewhere. A related attempt to construct a theoretical measure of the "net benefit" a country gains from foreign trade, analogous to the measures of consumer and producer surplus, was not entirely satisfactory as the partial equilibrium context had clearly been transcended. On matters of concrete trade policy for Britain, Marshall was a firm but cautious adherent of free trade, even unilateral free trade, but became increasingly concerned with the prospects for Britain's position in the world economy. The discussion in

Industry and Trade of the links between foreign competition and domestic industrial organization and structure reflected this concern.

I. Concept of National Income

It was Marshall who developed the concept of the national income, which he defined as follows. "The labour and capital of the country acting on its natural resources, produces annually a certain net aggregate of commodities-material and immaterial, including services of all kinds". This net output is called the national income. He believed that national income depends upon the co-operation of all the factors of production.

II. Agents of Production

Marshall distinguished between land, labour and capital. He held that capital is the result of man working upon natural resources. Hence there are only two essential agents (factors) of production- man (labour) and nature (land).While discussing the role played by the different agents of in production, Marshall pointed out that these factors were both competitiveand complementary. Though they competed for employment during theprocess of production in the short run, in the long run all these factors would complement one another.

III. Theory of Distribution

In the field of distribution Marshallmade use of the forces of supply and demand. He held that each factorwould be used in production to the extent where its marginal productivitywas equal to its marginal cost. The marginal productivity determinedthe demand price of the factor and the marginal cost determined itssupply price. However, this is not applicable in the determination of land rent.

IV. Prime Cost and Supplementary Cost

Marshall divided cost into prime costs and supplementary cost. Prime cost, which is now called variable cost, includes things like labour wages and raw materials. This cost changes in the short run along with production. A supplementary cost, which is now called fixed cost, include things like depreciation of the plant, interest on loans, rent, salaries of top executives etc. This do not change in the short run.

V. Principle of Substitution

In his analysis of marginal cost in relation to value, Marshall restated the principle of substitution. He stated that in the modern world production has become very complicated and nearly all the factors of production is controlled by businessmen and employers who had acquired specialized knowledge in organizing different economic processes. Then, the employer will choose those factors which are profitable to him. In other words, the employer will always substitute a costly factor by a less costly factor. This principle is also applicable to methods of production. If there are two methods of producing a commodity -by skilled and unskilled, then efficient in production to its cost.

VI. Monetary Theories

Marshall has made some original contributions to monetary economics. Marshall held that value of money is a function of its supply and demand for it. He put forward the 'purchasing power-parity' theory to explain the rate of exchange between countries with mutually inconvertible currencies. Besides, Marshall introduced the chain method of index numbers, proposed paper currency based on gold and silver symmetallism and distinguished between real rate of interest and money rate of interest.

16.2. The Aggregative Strand of Marshall's Thought

Though Marshall's attention was primarily directed to micro-economic problems, aggregative themes still occupied a place in his thought. In his view, the major micro-economic question was the determination of the general price level. Short term fluctuations in output and employment were peripheral matters; when they occurred they were expected to be temporary and slight. His analysis of the general price level was developed around a version of the 'quantity theory' of money. Much of the earlier discussion of this point of doctrine had proceeded from the tautological statement that the quantity of money multiplied by the number of times it was spent in a given time period (the velocity of circulation) would necessarily be equal to the average price level multiplied by the total number of transactions; this expression, after all, amounted to no more than two ways of viewing total expenditure. Marshall modified this procedure by shifting the focus from the rate at which the money supply turned over to an examination of the money balances held by the community. In the hands of one of Marshall's pupils this way of viewing money was later to open up fresh analytical horizons. Marshall's own results from the use of this 'cash balance' approach, however, were essentially no different from those that had been reached via the 'velocity of circulation' route. He maintained that the amount of money held was regulated by the institutional arrangements of the economy and, on ceteris paribus assumptions could be treated as a constant.

The effect of this procedure was to reinforce the essential requirement of Say's Law: that all income would be spent. The possibility of leakages into idle balances could, for practical purposes, be ignored. Money was interesting primarily in relation to spending and to the general price level, rather than for any connexion it might bear to the level of interest rates. This conclusion, of course, gained additional strength from Marshall's insistence - which was common to the neoclassical tradition as a whole - that interest rates would be established through the interaction of the supply of loanable funds (fed by saving) and the demand for loanable funds (stimulated by the productivity of capital). Moreover, the rate of interest could be relied upon to produce an equilibrium between decisions to save and to invest. Should the demand for loanable funds increase, the rate of interest would rise, making it more for people to reduce consumption spending and to save. Conversely; should the public choose to save more, the rate of interest would fall. Investors would then be induced to increase both their borrowings and their expenditures on plant and equipment. Further, this way of looking at the matter implied that the intersection of the curves of supply and demand for loanable funds determined the equilibrium rate of interest. The position of these curves, in turn, was established by the thriftiness of the community (on the supply side) and by the productivity of capital (on the demand side). This line of reasoning, while supporting a Say's Law interpretation of aggregative economic activity, did not preclude the possibility of economic instability. Though no disturbances of the scale experienced in the 1930s clouded the horizons of Marshall and his neoclassical contemporaries, they did observe modest cycles of boom and bust. How were these fluctuations to be explained? In Marshall's view the main answer was to be found in the psychology of the business community. Waves of optimism and pessimism seemed endemic to it. When business men were bullish the demand for loans increased. This phase might generate capital spending on many high risk undertakings, some of which were doomed to failure. And when they did fail the bubble was pricked. Pessimism replaced optimism as the dominant mood; investment and economic activity generally would be curtailed.

As Marshall described the process: The recent history of fluctuations of general credit shows much variety of detail, but a close uniformity of general outline. In the ascending phase, credit has been given somewhat boldly, and even to men whose business capacity has not been proved. For, at such times a man may gain a profit on nearly every transaction, even though he has brought no special knowledge or ability to bear on it; and his success may probably tempt others of like capacity with himself, to buy speculatively. If he is quick to get out of his ventures, he probably makes a profit. But his sales hasten a fall of prices, which must have come in the course of time. Though the fall is likely to be slight at first; yet each downward movement impairs the confidence which had caused the rise of prices, and is still giving them some support. The fall of a lighted match on something that smoulders has often started a disastrous panic in a crowded theatre. Credit cycles, however, were still incapable of converting 'partial' over-production into 'general' over production. Given time, the economic system would adjust itself to its normal full employment level of operations. No special action on the part of government was required to accomplish this result and, indeed, direct government intervention might make matters worse. The tendency toward instability could, however, be moderated by anticipatory action on the part of the monetary authorities whose proper role was to minimize discrepancies between prevailing interest rates and the rate that would be established through the normal interplay of the supply of and demand for loanable funds.

The main thrust of Marshall's aggregative analysis thus buttressed faith in the capability of the economic system, if left alone, to eradicate involuntary idleness. In the final edition (1920) of his principles, however, Marshall added one dark hint that the analytical basis for his conclusion might ultimately need to be revised. Following an orthodox neoclassical discussion of the relationship between productivity, thrift, and the rate interest, he inserted a note of qualification, everyone understands generally the causes which have kept the supply of accumulated wealth so small relatively to the demand for its use, that that use is on balance a source of gain, and can therefore require payment when loaned. Everyone is aware that the accumulation of wealth is held in check, and the rate of interest so far sustained, by the preference which the great mass of humanity have for present over deferred gratifications, or, in other words, by their unwillingness to wait. And indeed the true work of economic analysis in this respect is, not to emphasize this familiar truth, but to point out how much more numerous are the exceptions to this general preference than would appear at first sight.

16.3. Marshall and Economic Policy

By his own account, Marshall was originally attracted to the serious study of economics by a desire to understand the causes of poverty and the means by which it could be alleviated. He emerged from his investigation convinced that: the social and economic forces already at work are changing the distribution of wealth for the better: that they are persistent and increasing in strength; and that their influence is for the greater part cumulative; that the socio-economic organism is more delicate and complex than at first sight appears; and that large ill-considered changes might result in grave disaster. His sympathy for the sufferings of the mass of humanity had by no means diminished. But these impulses were now substantially tempered by the belief that radical measures to alter the existing economic order would be unwise. In particular, he opposed a socialistic programme on the grounds that: the collective ownership of the means of production would deaden the energies of mankind, and arrest economic progress; unless before its introduction the whole people had acquired a power of unselfish devotion to the public good which is now relatively rare. And, it might probably destroy much that is most beautiful and joyful in the private and domestic relations of life. These are the main, reasons which cause patient students of economics generally to anticipate little good and much evil from schemes for sudden and violent reorganization of the economic, social and political conditions of life. While the market system as portrayed by Marshall was largely benevolent, his analysis also demonstrated that in certain situations unregulated markets could not be relied upon to yield socially desirable results. Prominent among the exceptions were the cases in which - for technical reasons - competition would prove to be wasteful and inefficient, if not a practical impossibility. The 'natural monopolies' (a term which Marshall associated primarily with such public services as water supply, power generation, etc.) could not usefully be organized in accordance with the competitive plan and the case for government regulation (if not public ownership) in these instances was clear. He was reluctant, however, to recommend government intervention in those sectors of the economy in which increasing returns to scale threatened to produce industrial concentrations, even though these circumstances implied that individual firms could enjoy considerable market power and that prices would not be competitively determined. This problem, he contended, deserved continuing study. His general position on the life cycle of firms led him to the conclusion that the potential market power of large business units was unlikely to be abused for long. Though he was disposed to view the market as a sensitive instrument through which an economy's resources could be efficiently allocated, he also recognized that its performance could be improved. For this purpose, improvements in public education were particularly important. Consumers and producers could then conduct their affairs more intelligently by enhancing the rationality of their choices. Moreover, improved economic education would do much to eradicate one of the blights of an unregulated market system - the bouts of speculation which gave rise to harmful fluctuations.

Marshall was also prepared to entertain the possibility that governments could play a useful role in improving the allocative efficiency of markets. Would not the sum of social satisfactions be increased, he asked, if the productive resources of society were shifted in favour of lines of production subject to increasing returns and away from those in which decreasing returns prevailed? Greater outputs could then be obtained from the existing stock of resources. Governments could encourage a re-allocation of resources along these lines through appropriate taxes and subsidies. He advanced this suggestion, however, with the utmost caution, pointing out that such policies could be justified only when it could be shown that the gains in satisfactions arising from expanded outputs in the subsidized sectors more than offset the losses in utility associated with higher tax levies on others. He recognized that this criterion would be difficult to apply to practical cases.

Conceivably the introduction of maximization of aggregate utility as a goal of public policy could also be used to support recommendations for a redistribution of income. If it could be assumed that the marginal utility of money was likely to be greater far a poor man than for a rich one, it followed that society's aggregate satisfactions would be enlarged through redistribution from rich to poor. Marshall did not draw this conclusion. He did recommend a less systematic scheme of income redistribution for further study when he wrote of the possibilities of 'economic chivalry'. Such a regime would tax the rich to ameliorate the distress of those still trapped in poverty.

16.4. Keynesian Economics

With the recent economic crisis, there has been much talk of John Maynard Keynes and his economics. Keynes figured out the causes of the Great Depression and in doing so revolutionized the field of economics. According to Keynes, classical economics (basically that summarized by Alfred Marshall, including Ricardo, Mill, Edgeworth, and Pigou) addresses only a special case of the economy, while his is a more general theory. The classical theory of employment says the labour market is just another market: people get paid what they make and people only work if they get paid enough to make it worth it. Since it seemed unlikely that society had run out of money-making jobs, it was assumed that unemployment was caused either by people not knowing where the jobs were (frictional unemployment) or insisting on being paid more than they could make (voluntary unemployment). The classicals believed in Say's Law which state that supply creates its own demand. If there are people around willing to work, jobs will spring up to make use of them. If people are unemployed, it must be because they're refusing to take the job. We may explain some of the tenets of Keynesian economics. Keynesian economics is the view that in the short run, output is influenced by aggregate demand (total spending in the economy). In the Keynesian view, aggregate demand does not necessarily equal the productive capacity of the economy. Keynes contrasted his approach to the 'classical' economics that preceded his book. Keynesian economists argue that private sector decisions sometimes lead to inefficient macroeconomic outcomes. So it requires active policy (fiscal policy) by the government, in order to stabilize output over the business cycle. Keynesian economics advocates a mixed economy – predominantly private sector, but with a role for government intervention during recessions. The advent of the global financial crisis in 2008 has caused resurgence in Keynesian thought. Prior to the publication of Keynes's General Theory, mainstream economic thought was that the economy existed in a state of general equilibrium. This perception is reflected in Say's Law. Keynes's theory was significant because it reversed the mainstream thought of the time and brought about a greater awareness that problems such as unemployment are not a product of laziness, but the result of a structural inadequacy in the economic system. He argued that there was no guarantee that the goods that individuals produce would be met with demand and unemployment was a natural consequence. He saw the economy as unable to maintain itself at full employment and believed that it was necessary for the government to step in and put under-utilized savings to work through government spending. Keynesians therefore advocate an active stabilization policy to reduce the amplitude of the business cycle. According to the theory, government spending can be used to increase aggregate demand, thus increasing economic activity, reducing unemployment and deflation. The most important theory of Keynes was the theory of employment. Keynes criticized and rejected the classical theory of employment which denied the existence of unemployment or overproduction in the economy. The starting point of Keynesian theory of employment is the concept of effective demand. According to Keynes employment depends on effective demand and hence unemployment is due to lack of effective demand.

17
Rate of Interest

17.1. Rate of Interest

According to Keynes, interest is the reward for parting with liquidity. Liquidity preference is the preference of the people to keep their assets in the form of money. He argued that saving and investment are not the main determinants of interest rates, especially in the short run. According to him, the rate of interest is determined by the supply of money (Central Bank of a Country) and the demand for money (liquidity preference due to transaction, precautionary and speculative motives). Keynes developed the liquidity preference theory of interest according to which liquidity preference and supply of money determines the rate of interest.

(a) Fiscal Policy

Classical economists have traditionally yearned for balanced government budgets. Keynesians, on the other hand, believe this would aggravate the underlying problem. Keynes's view was that insufficient buying-power caused the depression. Keynes developed a theory which suggested that active government policy could be effective in managing the economy. Keynes advocated countercyclical fiscal policies, that is, policies that acted against the tide of the business cycle. That is deficit spending when a nation's economy suffers from recession or unemployment is persistently high– and the suppression of inflation in boom times by either increasing taxes or cutting back on government outlays. He argued that governments should solve problems in the short run rather than waiting for market forces to do it in the long run, because, "in the long run, we are all dead."Classical and Keynesian economics are two important schools of thought. Each school takes a different approach to the economic study of monetary policy, consumer behavior and government spending. A few basic distinctions separate these two schools.

(b) Basic Theory

Classical economic theory is rooted in the concept of a laissez-faire economic market. A laissez-faire– also known as free–market requires little to no government intervention. It also allows individuals to act according to their own self interest regarding economic decisions. This ensures economic resources are allocated according to the desires of individuals and businesses in the marketplace. Keynesian economic theory relies on spending and aggregate demand. Keynesian economists believe the aggregate demand is influenced by public and private decisions. Public decisions represent government agencies and Private decisions include individuals and Government spending is not a major force in a classical economic theory. Classical economists believe that consumer spending and business investment represents the more important parts of a nation and economic growth. Too much government spending takes away valuable economic resources needed by individuals and businesses. To classical economists, government spending and involvement can retard a nation's economic growth. Keynesian economics relies on government spending to a nation's economic growth during sluggish economic downturns. Similar to classical economists, Keynesians believe the nation's economy is made up of consumer spending, business investment and government spending. However, Keynesian theory dictates that government spending can improve economic growth in the absence of consumer spending or business investment.

(c) Short versus Long Run

Classical economics focuses on creating long-term solutions for economic problems. The effects of inflation, government regulation and taxes can all play an important part in developing classical economic theories. Classical economists also take into account the effects of other current policies and how new economic theory will improve or distort the free market environment. Keynesian economics often focuses on immediate results in economic theories. Policies focus on the short-term needs and how economic policies can make instant corrections to a nation's economy. This is why government spending is such a key component of Keynesian economics. During economic recessions and depressions, individuals and businesses do not usually have the resources for creating immediate results through consumer spending or business investment. The government is seen as the only force to end these downturns through fiscal policies providing instant economic results. Classicals follow the basic assumption that economy is in full employment. The wages and prices are very flexible and there is no need of fiscal or monetary policy. The invisible hand makes the economy self correctable. The Aggregate supply curve is vertical according to Classicals and so any rise in aggregate demand will increase prices and not production. In Classical Theory, Government has minimal role in the economy, and the macro-economy is self adjusting; meaning consumers and businesses will correct any problems with the economy automatically over time. Classical theory focuses on long-term goals. Keynesian economics follow the basic assumptions that Economy may not be in full employment in the short run. Wages are rigid and prices are sticky. Fiscal policy may be needed to correct the disequilibrium or improve the efficiency of the economy. Aggregate supply is upward sloping in the short run so a rise in aggregate demand may raise the production. In Keynesian theory, Government has a large role in the economy, and focuses on short-term goals.

17.2. Keynes as a Critic of Classical Economics

Classical economists generally include Adam Smith, Jeremy Bentham, T.R.Malthus, David Ricardo, J.S.Mill, J.B.Say etc. But Keynes included, the neoclassical economists including Marshall, Pigou and Edgeworth in the category of classical economists. Hence, Keynes criticised the economic theories developed by these writes mainly Pigou. The main points of contrast between the Keynesian and Classical theories of Income and Employment are discussed in brief as under. The classical economists explained unemployment using traditional partial equilibrium supply and demand analysis. According to them unemployment results when there is an excess supply of labour at a particular higher wage level. By accepting lower wage, the unemployed workers will go back to their jobs and the equilibrium between demand for labour and supply of labour will be established in the labour market in the long period. This equilibrium in the economy is always associated with full employment level. According to classical economists unemployment results when the wage level of workers is above the equilibrium wage level and as a result there of, the quantity of labour supplied is higher than quantity of labour demanded. The difference between the two (supply and demand) is unemployment. J.M. Keynes and his followers, however reject the fundamental classical theory of full employment equilibrium in the economy. They consider it as unrealistic. According to them full employment is a rare phenomenon in the capitalistic economy. The unemployment occurs they say when the aggregate demand function intersects the aggregate supply function at a point of less than full employment level. Keynes suggested that in the short period, the government can raise aggregate demand in the economy through public investment programmes to reduce unemployment.

17.2.1. Say's Law of Market

Say's Law "Supply creates its own demand" is central to the classic vision of the economy.

According to say the production of goods and services generates expenditure sufficient to ensure that they are sold in the market. There is no deficiency of demand for goods and hence no need to unemployed workers. According to him full employment is a normal condition of market economy. J.M. Keynes has strongly refuted. Say's Law of Market with the help of effective demand. Effective demand is the level of aggregate demand which is equal to aggregate supply. Whenever there is efficiency in aggregate demand a part of the goods produced remain unsold in the market which lead to general over production of goods and services in the market. When all the goods produced in the market are not sold, the firms lay off workers. The deficiency in demand for goods creates unemployment in the economy. Classical economists are of the view that saving and investment are equal at the full employment level. If at any time flow of savings is greater than flow of investment, then the rate of interest declines in the money market this leads to an increase in investment. The process continues till the flow of investment equals the flow of saving. Thus according to the classical economists, the quality between saving and investment is brought about through mechanism of rate of interest. J.M.Keynes is however of the view that equality between saving and investment is brought about through changes income rather than the changes in interest rate.

17.2.2. Money and Prices

The classical economists are of the opinion that price level varies in response to changes in the quantity of money. The quantity theory of money seeks to explain the value of money in terms of changes in its quantity. J.M.Keynes has rejected the simple quantity theory of money. According to him if there is recession in the economy and the resources are lying idle and unutilized an increased spending of money lead to substantial increase in real output and employment without affectively price level. There are some similarities and more dissimilarity between Marx and Keynes. Both of them were interested in macroeconomics. Both of them regarded inadequate demand as the cause of trade cycles. Both emphasized the falling rate of profit. Both believed that the rewards for capitalist groups were excessive. Keynes thought that capitalist system could be preserved through the modifications of the capitalist institutions and changes in policy, Marx predicted the inevitable decline and fall of capitalism. While Marx became the 'prophet of doom', Keynes became the 'prophet of boom'. Keynes was the very opposite of Marx and Keynesian philosophy should be regarded as a major alternative to Marxism. Of course, Keynes conceded that Marx had some idea of the unemployment problem in a capitalist society.

New classical economics introduced a set of macroeconomic theories that were based on optimising microeconomic behaviour. These models have been developed into the Real Business Cycle Theory, which argues that business cycle fluctuations can to a large extent be accounted for by real (in contrast to nominal) shocks. Beginning in the late 1950s new classical macroeconomists began to disagree with the methodology employed by Keynes and his successors. New classical theorists demanded that macroeconomics be grounded on the same foundations as microeconomic theory, profit-maximizing firms and rational, utility-maximizing consumers. Monetarism is a school of economic thought that emphasizes the role of monetary authority (Central Bank) in controlling the amount of money in circulation for stability and growth. It is the view within monetary economics that variation in the money supply has major influences on national output in the short run and the price level over longer periods and those objectives of monetary policy are best met by targeting the growth rate of the money. Monetarism today is mainly associated with the work of Milton Friedman, who was among the generation of economists to accept Keynesian economics and then criticize Keynes' theory of gluts based on a policy of government intervention. Thus, Monetarism is an economic theory which focuses on the macroeconomic effects of the supply of money and central banking. It argues that excessive expansion of the money supply is inherently inflationary, and that monetary authorities should focus solely on maintaining price stability.

Friedman and Anna Schwartz wrote an influential book, A Monetary History of the United States, 1867- 1960, and argued that "inflation is always and everywhere a monetary phenomenon." It attributed deflationary spirals to the reverse effect of a failure of a central bank to support the money supply during a liquidity crunch. Friedman advocated, a central bank policy aimed at keeping the supply and demand for money at equilibrium, as measured by growth in productivity and demand. Friedman focused on price stability, which is the equilibrium between supply and demand for money. The rise of monetarism accelerated from Milton Friedman's 1956 restatement of the quantity theory of money. Friedman argued that the money could be described as depending on a small number of economic variables. Thus, where the money supply expanded, people would not simply wish to hold the extra money in idle money balances; *i.e.*, if they were in equilibrium before the increase, they were already holding money balances to suit their requirements, and thus after the increase they would have money balances surplus to their requirements. These excess money balances would therefore be spent and hence aggregate demand would rise. Similarly, if the money supply were reduced people would want to replenish their holdings of money by reducing their spending. In this, Friedman challenged a simplification attributed to Keynes suggesting that "money does not matter." Thus the word 'monetarist' was coined. Milton Friedman and Anna Schwartz in their book A Monetary History of the United States, 1867-1960 argued that the Great Depression of 1930 was caused by a massive contraction of the money supply and not by the lack of investment Keynes had argued. They also maintained that post-war inflation was caused by an over-expansion of the money supply. Monetarists of the Milton Friedman school of thought believed in the 1970s and 1980s that the growth of the money supply should be based on certain formulations related to economic growth. As such, they can be regarded as advocates of a monetary policy based on a "quantity of money" target. There was debate between Monetarists and Keynesians in the 1960s over the role of government in stabilizing the economy. Both Monetarists and Keynesians are in agreement over the fact that issues such as business cycles, unemployment, inflation are caused by inadequate demand, and need to be addressed, but they had fundamentally different perspectives on the capacity of the economy to find its own equilibrium and as a consequence the degree of government intervention that is required to create equilibrium. Keynesians emphasized the use of discretionary fiscal policy and monetary policy, while monetarists argued the primacy of monetary policy, and that it should be rules based. The debate was largely resolved in the 1980s. Since then, economists have largely agreed that central banks should bear the primary responsibility for stabilizing the economy, and that monetary policy should largely follow the Taylor rule – which many economists credit with the Great Moderation. The Global Financial Crisis, however, has convinced many economists and governments of the need for fiscal interventions and highlighted the difficulty in stimulating economies through monetary policy alone during a liquidity trap.

17.3. Rational Expectations School

Rational expectations is a hypothesis in economics which states that agents' predictions of the future value of economically relevant variables are not systematically wrong in that all errors are random. Equivalently, this is to say that agents' expectations equal true statistical expected values. An alternative formulation is that rational expectations are model-consistent expectations, in that the agents inside the model assume the model's predictions are valid. The rational expectations assumption is used in many contemporary macroeconomic models, game theory and applications of rational choice theory. Since most macroeconomic models today study decisions over many periods, the expectations of workers, consumers and firms about future economic conditions are an essential part of the model. How to model these expectations has long been controversial, and it is well known that the macroeconomic predictions of the model may differ depending on the assumptions made about expectations To assume rational expectations is to assume that agents' expectations may be individually wrong, but are correct *on average*. In other words, although the future is not fully predictable, agents' expectations are assumed not to be systematically biased and use all relevant information in forming expectations of economic variables. This way of modeling expectations was originally proposed by John F. Muth and later became influential when it was used by Robert Lucas, Jr. and others. Modeling expectations is crucial in all models which study how a large number of individuals, firms and organizations make choices under uncertainty. For example, negotiations between workers and firms will be influenced by the expected level of inflation, and the value of a share of stock is dependent on the expected future income from that stock. Rational expectations theory defines this kind of expectations as being identical to the best guess of the future (the optimal forecast) that uses all available information. Thus, it is assumed that outcomes that are being forecast do not differ systematically from the market equilibrium results. As a result, rational expectations do not differ systematically or predictably from equilibrium results. That is, it assumes that people do not make systematic errors when predicting the future, and deviations from perfect foresight are only random. In an economic model, this is typically modeled by assuming that the expected value of a variable is equal to the expected value predicted by the model. For example, suppose that P is the equilibrium price in a simple market, determined by supply and demand.

The theory of rational expectations says that the actual price will only deviate from the expectation if there is an 'information shock' caused by information unforeseeable at the time expectations were formed. Rational expectations theories were developed in response to perceived flaws in theories based on adaptive expectations. Under adaptive expectations, expectations of the future value of an economic variable are based on past values. For example, people would be assumed to predict inflation by looking at inflation last year and in previous years. Under adaptive expectations, if the economy suffers from constantly rising inflation rates (perhaps due to government policies), people would be assumed to always underestimate inflation. This may be regarded as unrealistic - surely rational individuals would sooner or later realize the trend and take it into account in forming their expectations.

The hypothesis of rational expectations addresses this criticism by assuming that individuals take all available information into account in forming expectations. Though expectations may turn out incorrect, they will not deviate systematically from the expected values. The rational expectations hypothesis has been used to support some radical conclusions about economic policymaking. An example is the

Proposition developed by Thomas Sargent and Neil Wallace. If the Federal Reserve attempts to lower unemployment through expansionary monetary policy economic agents will anticipate the effects of the change of policy and raise their expectations of future inflation accordingly. This in turn will counteract the expansionary effect of the increased money supply. All that the government can do is raise the inflation rate, not employment. This is a distinctly New Classical outcome. During the 1970s rational expectations appeared to have made previous macroeconomic theory largely obsolete, which culminated with the Lucas critique. However, rational expectations theory has been widely adopted throughout modern macroeconomics as a modeling assumption thanks to the work of New Keynesians such as Stanley Fischer. Rational expectations theory is the basis for the efficient market hypothesis (efficient market theory). If a security's price does not reflect all the information about it, then there exist "unexploited profit opportunities": someone can buy (or sell) the security to make a profit, thus driving the price toward equilibrium. In the strongest versions of these theories, where all profit opportunities *have been* exploited, all prices in financial markets are correct and reflect market fundamentals (such as future streams of profits and dividends). Each financial investment is as good as any other, while a security's price reflects all information about its intrinsic value.

17.4. Criticisms of Rational Expectations School

The models of Muth and Lucas (and the strongest version of the efficient-market hypothesis) assume that at any specific time, a market or the economy has only one equilibrium (which was determined ahead of time), so that people form their expectations around this unique equilibrium. Muth's math (sketched above) assumed that P^* was unique. Lucas assumed that equilibrium corresponded to a unique "full employment" level (potential output) – corresponding to a unique NAIRU or natural rate of unemployment. If there is more than one possible equilibrium at any time then the more interesting implications of the theory of rational expectations do not apply. In fact, expectations would determine the nature of the equilibrium attained, reversing the line of causation posited by rational expectations theorists. A further problem relates to the application of the rational expectations hypothesis to aggregate behavior. It is well known that assumptions about individual behavior do not carry over to aggregate behavior. The same holds true for rationality assumptions: Even if all individuals have rational expectations, the representative household describing these behaviors may exhibit behavior that does not satisfy rationality assumptions (Janssen, 1993). Hence the rational expectations hypothesis, as applied to the representative household, is unrelated to the presence or absence of rational expectations on the micro level and lacks, in this sense, a microeconomic foundation.

It can be argued that it is difficult to apply the standard efficient market hypothesis (efficient market theory) to understand the stock market bubble that ended in 2000 and collapsed thereafter; however, advocates of rational expectations say that the problem of ascertaining all the pertinent effects of the stock-market crash is a great challenge. Those studying financial markets similarly apply the efficient markets hypothesis but keep the existence of exceptions in mind.

17.4.1. Supply Side Economics

Supply-side economics developed during the 1970s in response to Keynesian economic policy, and in particular the failure of demand management to stabilize Western economies during the stagflation of the 1970s, in the wake of the oil crisis in 1973. It drew on a range of non-Keynesian economic thought, particularly the Chicago School and Neoclassical School. Supply-side economics is a school of macroeconomics that argues that economic growth can be most effectively created by lowering barriers for people to produce (supply) goods and services, such as lowering income tax and capital gains tax rates, and by allowing greater flexibility by reducing regulation. According to supply-side economics, consumers will then benefit from a greater supply of goods and services at lower prices. Policy recommendations of supply-side economists are lower marginal tax rates and less regulation. The Laffer curve embodies a tenet of supply side economics: those government tax revenues from a specific tax are the same (nil) at 100% tax rates as at 0% tax rates respectively. The tax rate that achieves optimum, or highest government revenues is somewhere in between these two values. What most separates supply-side economics as a modern phenomenon is its argument in favor of a low tax rate. Thus, Supply-side economists argued for tax cuts for increasing supply and economic growth. As in classical economics, supply-side economics proposed that production or supply is the key to economic prosperity and that consumption or demand is merely a secondary consequence.

17.4.2. New Keynesian Economics

New Keynesian Economics Is A School of Contemporary macroeconomics that strives to provide microeconomic foundations for Keynesian economics. It developed partly as a response to criticisms of Keynesian macroeconomics by adherents of New Classical macroeconomics. Two main assumptions define the New Keynesian approach to macroeconomics. Like the New Classical approach, New Keynesian macroeconomic analysis usually assumes that households and firms have rational expectations. But the two schools differ in that New Keynesian analysis usually assumes a variety of market failures. In particular, New Keynesians assume that there is imperfect competition in price and wage setting to help explain why prices and wages can become "sticky", which means they do not adjust instantaneously to changes in economic conditions. Wage and price stickiness, and the other market failures present in New Keynesian models, imply that the economy may fail to attain full employment. Therefore, New Keynesians argue that macroeconomic stabilization by the government (using fiscal policy) or by the central bank (using monetary policy) can lead to a more efficient macroeconomic outcome than a laissez faire policy would.

17.4.3. Neo-liberalism

Neo-liberalism is a political philosophy whose advocate support economic liberalization, free trade open markets, privatisation, deregulation, and decreasing the size of the public sector while increasing the role of the private sector in modern society. The term was introduced in the late 1930s by European liberal scholars to promote a new form of liberalism after interest in liberalism had declined in

Europe. In the decades that followed, neoliberal theory tended to be at variance with the more laissez-faire doctrine of classical liberalism and promoted instead a market economy under the guidance and rules of a strong state, a model which came to be known as the social. In the sixties, usage of the term "neoliberal" heavily declined. When the term was reintroduced in the following decades, the meaning had shifted. The term neoliberal is now normally associated with laissez-faire economic policies, and is used mainly by those who are critical of legislative market reform. The term "neoliberalism" was originally coined in 1938 by the German scholar Alexander Rustow. It is defined as "the priority of the price mechanism, the free enterprise, the system of competition and a strong and impartial state." To be "neoliberal" meant that – in the name of liberalism – a modern economic policy is required. Neoliberalism was not a monolithic theory. At the outset it drew on different academic approaches such as the the Austrian School, the Chicago School of Economics etc.

The term neoliberalism is almost never defined but used in several senses to describe ideology, economic theory, development theory, or economic reform policy. And it now suggests a market fundamentalism closer to the Laissez-Faire principle. This leaves some controversy as to the precise meaning of the term and its usefulness as a descriptor in the social sciences, especially as the number of different kinds of market economies have proliferated in recent years. Now a day the most common use of the term neoliberalism refers to economic reform policies such as "eliminating price controls, deregulating capital markets and lowering trade barriers", and reducing state influence on the economy especially by privatization and fiscal austerity. The term is used in several senses: as a development model it refers to the rejection of structuralist economics in favour of the Washington Consensus; as an ideology the term is used to denote a conception of freedom as an overarching social value associated with reducing state functions to those of a minimal state; and finally as an academic paradigm the term is closely related to neoclassical economic theory. Some people believe that the term is used as a pejorative for policies that deregulate the private sector and increase its role in the economy.

Dependency theory originates with two papers published in 1949 – one by Hans Singer, one by Raul Prebisch – in which the authors observe that the terms of trade for underdeveloped countries relative to the developed countries had deteriorated over time: the underdeveloped countries were able to purchase fewer and fewer manufactured goods from the developed countries in exchange for a given quantity of their raw materials exports. Dependency theory is a body of social science theories predicated on the notion that resources flow from a "periphery" of poor and underdeveloped states to a "core" of wealthy states, enriching the latter at the expense of the former. It is a central contention of dependency theory that poor states are impoverished and rich ones enriched by the way poor states are integrated into the "world system." The theory arose as a reaction to modernization theory, an earlier theory of development which held that all societies progress through similar stages of development, that today's underdeveloped areas are thus in a similar situation to that of today's developed areas at some time in the past, and that therefore the task in helping the underdeveloped areas out of poverty is to accelerate them along this supposed common path of development, by various means such as investment, technology transfers, and closer integration into the world market. Dependency theory rejected this view, arguing that underdeveloped countries are not merely primitive versions of developed countries, but has unique features and structures of their own; and, importantly, is in the situation of being the weaker members in a world market economy.

The premises of dependency theory are that: Poor nations provide natural resources, cheap labour, a destination for obsolete technology, and markets for developed nations, without which the latter could not have the standard of living they enjoy. Wealthy nations actively perpetuate a state of dependence by various means. This influence may be multifaceted, involving economics, media control, politics, banking and finance, education, culture, sport, and all aspects of human resource development (including recruitment and training of workers). Wealthy nations actively counter attempts by dependent nations to resist their influences by means of economic sanctions and/or the use of military force. Dependency theory states that the poverty of the countries in the periphery is not because they are not integrated into the world system, or not 'fully' integrated as is often argued by free market economists, but because of how they are integrated into the system. This introduces a paradoxical effect, in that although both the first and third-world countries are benefitting, the poorer side is being locked into detrimental economic position. They rely on the rich for the little work that is available to them, yet this causes a barrier from the nation growing independently. In a future perspective, such nations have no opportunity to improve their quality of life.

18
Economics of Caste System

18.1. Economics of Caste System

According to Ambedkar the caste system in India was a major obstacle to economic growth and development. The caste system didn't allow people to teach their professional skills to any person belonging to other caste. Only the members of their own caste were allowed to learn the profession. Thus if a person had the skill necessary for a particular job he would not accept the profession of a caste lower than his own. In a dynamic industrial set up the individual must be free to choose his occupation. But due to social religious restriction on inter occupational mobility has following consequences: Firstly, by not permitting readjustment of occupation, caste become a direct cause of much of the unemployment in various groups, as a religious Hindu would prefer to be unemployed rather than getting employed in profession not assigned to his caste. Second, individual justice and economic efficiency demand that competition exists in factor market. Due to the restriction on inter-occupational mobility of labour, capital and entrepreneurship across caste groups the caste system creates segregation in each of these markets. Labour and capital thus does not flow from one occupation to another even if the wage rate or rates of return on investments are higher in the alternative occupations.

This brings about a high level of inefficiency in resource allocation. The division of occupation is not based on individual choice. Some of the occupations are socially degrading, and people are forced in these occupations on account of their caste origin. Such people cannot derive job satisfaction. As economic organisation caste is therefore a harmful institution in as much as it involves the subordination of man's natural powers an inclination to the exigencies of social rules. This also results in disassociation of intelligence from work. The dignity of labour is nearly absent in the general scheme of theory of caste. The practice of untouchability has ruined the nation and the economy as a whole. Those belonged to the lower strata of society or the untouchables were not allowed to mix with others and denied basic civil right. They had no access to sources of income. They had no right to education. They had no right tom property and were prohibited from acquiring wealth. They could not use their labour for their own upliftment. The Caste system has crippled the untouchable as it denied them economic security and independence.

In the opinion of Ambedkar the economics of the caste system had six type s of consequences: 1) it divided labour ii) it disassociated work from interest iii) it disconnected intelligence from the manual labourer iv) it devitalized persons by denying them right to cultivate vital interest v) it prevented mobilization vi) it deprived Sudras of all economic avenues of employment and took them nearly to the position of a slave. Ambedkar characterised Gandhism as conservatism and purely imaginary. He remarked that Gandhism meant return to village and making the village self sufficient and as such it made Gandhism a mere matter of regionalism. Gandhiji vehemently opposed machinery. He idolized charkha (spinning wheel) and called western civilisation as the creation of satan. He even said "I would not weep over the disappearance of machinery or consider it a calamity". Ambedkar opposed Gandhism because, in his opinion, Gandhism was against democracy and also against mechanisation and civilization which were vital for progress of a country. According to Ambedkar, democratic society cannot be indifferent to machinery. Gandhism might well be suited to a society which did not accept democracy as its ideal. He maintained that the ills were not due to machinery and they were wrong due to social organisation which has made private property and pursuit of personal gain matters of absolute sanctity. He remarked that "Gandhism is a paradox. It seeks to maintain intact a social structure which permits the domination of one class by another on a hereditary basis which means a perpetual domination of one class by another."

Ambedkar didn't have faith in laissez faire or Trusteeship of Gandhiji. He believed in socialism and advocated state socialism to ensure proper working of national economy. He favoured land acquisition by the state for distribution among poor. In his view socialism will not come without directive principle. Ambedkar viewed, Gandhism, a philosophy of the well to do and the leisured class. It deluded people into accepting misfortunes by presenting them as best of good fortunes. Ambedkar's views on communism are presented in his essay "Buddhism and Communism". Like Marx Ambedkar believed that there is exploitation in the world, the poor are exploited by the rich; there is enslavement of the masses by the privileged few which leads to perpetuation of poverty and ultimately suffering. Ambedkar did not sympathise with the Marxist paradigm of development. Ambedkar was of the opinion that economic relationships are not everything in human life nor the economic motive was the sole driving power behind all human activity. He emphasised that exploitation has many dimensions like social, religious and political. In Ambedkar's opinion, communism advocates revolution, it recommends revolutionary methods for overcoming the opposition of the capitalists for establishment of the dictatorship of the proletariat and any action that serves the interest of the proletarian revolution is ethical and there is no consideration for human life. Ambedkar being an ardent follower of democracy and human right was against any form of dictatorship. He was not in favour of violent change rather he believed that constitutional provision and democratic means could bring desired reformation. The control of all means of production and the abolition of private property as advocated by Marx were not accepted by Ambedkar. Ambedkar was not in tune with Marxian convictions that abolition of private property would bring an end to the poverty and suffering of the have nots and that the state is a temporary institution which will wither away in course of time.

Ambedkar believed that state is fundamental to human existence. It has many important roles to play. He maintained that state would continue to exist as long as human society survived. He believed in classless society but not in a stateless society. He sought active involvement of the government in economic development. In his book, "States and Minorities", he stated that state has an obligation to plan the economic life of the people. Ambedkar's concept of state socialism is based on these tenets:

a) State ownership economy agricultural land and key industries
b) Maintenance of productive resources by the states
c) A just distribution of the common produce among different people without any discrimination of caste or creed.

Ambedkar was primarily an economist and his achievement in the field was primarily overshadowed by his achievement in his fields of politics. His political ideas had moral purpose. He was more concerned about the human welfare. He had unflinching faith in socialism and believed that it would solve the problem of the under privileged. Mixed economy was the cornerstone of his philosophy. He was concerned about the equitable distribution of wealth among people. He advocated the modernization of Indian society and economy. Ambedkar also linked salvation and emancipation of untouchable people with the emergence of modern urbanized and industrialized India with modern liberal ideology.

18.2. Internatoinal Trade Contents

International trade is the exchange of capital, goods, and services across international borders or territories. In most countries, such trade represents a significant share of gross domestic product (GDP). While international trade has been present throughout much of history, its economic, social, and political importance has been on the rise in recent centuries.

Industrialization, advanced transportation, globalization, multinational corporations, and outsourcing are all having a major impact on the international trade system. Increasing international trade is crucial to the continuance of globalization. Without international trade, nations would be limited to the goods and services produced within their own borders. International trade is, in principle, not different from domestic trade as the motivation and the behavior of parties involved in a trade do not change fundamentally regardless of whether trade is across a border or not. The main difference is that international trade is typically more costly than domestic trade. The reason is that a border typically imposes additional costs such as tariffs, time costs due to border delays and costs associated with country differences such as language, the legal system or culture.

Another difference between domestic and international trade is that factors of production such as capital and labour are typically more mobile within a country than across countries. Thus international trade is mostly restricted to trade in goods and services, and only to a lesser extent to trade in capital, labour or other factors of production. Trade in goods and services can serve as a substitute for trade in factors of production. Instead of importing a factor of production, a country can import goods that make intensive use of that factor of production and thus embody it. An example is the import of labour-intensive goods by the United States from China. Instead of importing Chinese labour, the United States imports goods that were produced with Chinese labour.

18.3. Models of International Trade

The following are noted models of international trade.

(a) Adam Smith's Model

Adam Smith displays trade taking place on the basis of countries exercising absolute advantage over one another.

(b) Ricardian Model

The law of comparative advantage was first proposed by David Ricardo. The Ricardian model focuses on comparative advantage, which arises due to differences in technology or natural resources. The Ricardian model does not directly consider factor endowments, such as the relative amounts of labour and capital within a country. The Ricardian model makes the following assumptions:

1. Labour is the only primary input to production
2. The relative ratios of labour at which the production of one good can be traded off for another differ between countries and governments

(c) Heckscher-Ohlin Model

In the early 1900s a theory of international trade was developed by two Swedish economists, Eli Heckscher and Bertil Ohlin. This theory has subsequently been known as the Heckscher-Ohlin model (H-O model). In the Heckscher-Ohlin model the pattern of international trade is determined by differences in factor endowments. It predicts that countries will export those goods that make intensive use of locally abundant factors and will import goods that make intensive use of factors that are locally scarce. Empirical problems with the H-O model, such as the Leontief paradox, were noted in empirical tests by Wassily Leontief who found that the United States tended to export labour-intensive goods despite having an abundance of capital.

The H-O model makes the following core assumptions:

1. Labour and capital flow freely between sectors
2. The amount of labour and capital in two countries differ (difference in endowments)

3. Technology is the same among countries (a long-term assumption)
4. Tastes are the same.

18.4. Reality and Applicability of the Heckscher-Ohlin Model

In 1953, Wassily Leontief published a study in which he tested the validity of the Heckscher-Ohlin theory. The study showed that the U.S was more abundant in capital compared to other countries; being more abundant in capital, U.S would be expected to export capital-intensive goods and import labour-intensive ones. Nonetheless Leontief found out that the U.S's exports were less capital intensive than its imports. This is what is referred to as Leontief's paradox.

After the appearance of Leontief's paradox, many researchers tried to save the Heckscher-Ohlin theory, either by new methods of measurement, or by new interpretations. Leamer emphasized that Leontief did not interpret H-O theory properly and claimed that with a right interpretation, the paradox did not occur. Brecher and Choudri found that, if Leamer was right, the American workers' consumption per head should be lower than the workers' world average consumption. Many textbook writers, including Krugman and Obstfeld and Bowen, Hollander and Viane, are negative about the validity of H-O model. After examining the long history of empirical research, Bowen, Hollander and Viane concluded: "Recent tests of the factor abundance theory [H-O theory and its developed form into many-commodity and many-factor case] that directly examine the H-O-V equations also indicate the rejection of the theory."

In the specific factors model, labour mobility among industries is possible while capital is assumed to be immobile in the short run. Thus, this model can be interpreted as a short-run version of the Heckscher-Ohlin model. The "specific factors" name refers to the assumption that in the short run, specific factors of production such as physical capital are not easily transferable between industries. The theory suggests that if there is an increase in the price of a good, the owners of the factor of production specific to that good will profit in real terms.

Additionally, owners of opposing specific factors of production (*i.e.*, labour and capital) are likely to have opposing agendas when lobbying for controls over immigration of labour. Conversely, both owners of capital and labour profit in real terms from an increase in the capital endowment. This model is ideal for understanding income distribution but awkward for discussing the pattern of trade

18.5. New Trade Theory

New Trade Theory tries to explain empirical elements of trade that comparative advantage-based models above have difficulty with. These include the fact that most trade is between countries with similar factor endowment and productivity levels, and the large amount of multinational production (*i.e.* foreign direct investment) that exists. New Trade theories are often based on assumptions such as monopolistic competition and increasing returns to scale. One result of these theories is the home-market effect, which asserts that, if an industry tends to cluster in one location because of returns to scale and if that industry faces high transportation costs, the industry will be located in the country with most of its demand, in order to minimize cost. Although new trade theory can explain the growing trend of trade volumes of intermediate goods, Krugman's explanation depends too much on the strict assumption that all firms are symmetrical, meaning that they all have the same production coefficients. Shiozawa, based on much more general model, succeeded in giving a new explanation on why the traded volume increases for intermediate goods when the transport cost decreases.

18.5.1. Gravity Model

The Gravity model of trade presents a more empirical analysis of trading patterns. The gravity model, in its basic form, predicts trade based on the distance between countries and the interaction of the countries' economic sizes. The model mimics the Newtonian law of gravity which also considers distance and physical size between two objects. The model has been proven to be empirically strong through econometric analysis. The Ricardian theory of comparative advantage became a basic constituent of neoclassical trade theory. Any undergraduate course in trade theory includes a presentation of Ricardo's example of a two-commodity, two-country model. A common representation of this model is made using an Edgeworth Box. This model has been expanded to many-country and many-commodity cases. Major general results were obtained by McKenzie and Jones, including his famous formula. It is a theorem about the possible trade pattern for N-country N-commodity cases. Ricardo's idea was even expanded to the case of continuum of goods by Dornbusch, Fischer, and Samuelson. This formulation is employed for example by Matsuyama and others. These theories use a special property that is applicable only for the two-country case. Inspired by Piero Sraffa, a new strand of trade theory emerged and was named neo-Ricardian trade theory. The main contributors include Ian Steedman (1941-) and Stanley Metcalfe (1946-). They have criticized neoclassical international trade theory, namely the Heckscher-Ohlin model on the basis that the notion of capital as primary factor has no method of measuring it before the determination of profit rate (thus trapped in a logical vicious circle). This was a second round of the Cambridge capital controversy, this time in the field of international trade.

18.5.2. Ricardo-Sraffa Trade Theory

John Chipman observed in his survey that McKenzie stumbled upon the questions of intermediate products and discovered that "introduction of trade in intermediate product necessitates a fundamental alteration in classical analysis." It took many years until Y. Shiozawa succeeded in removing this deficiency. The Ricardian trade theory was now constructed in a form to include intermediate input trade for the most general case of many countries and many goods. This new theory is called Ricardo-Sraffa trade theory.

Based on an idea of Takahiro Fujimoto, who is a specialist in automobile industry and a philosopher of the international competitiveness, Fujimoto and Shiozawa developed a discussion in which how the factories of the same multi-national firms compete between them across borders. International *intra-firm competition* reflects a really new aspect of international competition in the age of so-called **global competition**. International Production Fragmentation Trade Theory.

Fragmentation and International Trade Theory widens the scope for "application of Ricardian comparative advantage." In his chapter entitled *Li & Fung, Ltd.: An agent of global production* (2001), Cheng used Li & Fong Ltd as a case study in the international production fragmentation trade theory through which producers in different countries are allocated a specialized slice or segment of the value chain of the global production. Allocations are determined based on "technical feasibility" and the ability to keep the lowest final price possible for each product. An example of fragmentation theory in international trade is Li & Fung's garment sector network with yarn purchased in South Korea, woven and dyed in Taiwan, the fabric cut in Bangladesh, pieces assembled in Thailand and the final product sold in the United States and Europe to major brands.

In 1995 Li & Fung Ltd purchased Inchcape Buying Services, an established British trading company and widely expanded production in Asia. Li & Fung supplies dozens of major retailers, including Wal-Mart Stores, Inc., branded as Walmart.

Largest Countries by Total International Trade Rank	*Country*	*International Trade of Goods (Billions of USD)*	*Date of Information*
-	World	36,534.0	2012 est.
-	**European Union**	4,468.6	2012 est.
1	**United States**	3,882.4	2012 est.
2	**China**	3,866.9	2012 est.
3	**Germany**	2,575.7	2012 est.
4	**Japan**	1,684.4	2012 est.
5	**Netherlands**	1,247.4	2012 est.
6	**France**	1,243.9	2012 est.
7	**United Kingdom**	1,149.4	2012 est.
8	**South Korea**	1,067.5	2012 est.
9	**Italy**	986.9	2012 est.
10	**Hong Kong**	947.9	2012 est.
11	**Canada**	917.3	2012 est.
12	**Belgium**	882.0	2012 est.
13	**Russia**	864.7	2012 est.
14	**Singapore**	788.1	2012 est.
15	**India**	779.4	2012 est.
16	**Mexico**	751.4	2012 est.
17	**Spain**	624.9	2012 est.
18	**Taiwan**	571.8	2012 est.
19	**Saudi Arabia**	529.8	2012 est.
20	**Australia**	518.2	2012 est.

19
Economic Ideas of St. Thomas Aquinas

19.1 Economic Ideas of St. Thomas Aquinas

Saint Thomas Aquinas (1225–1274) was an Italian Catholic Priest and one of the most important medieval philosophers and theologians. He was immensely influenced by scholasticism and Aristotle and known for his synthesis of the two aforementioned traditions. Although he wrote many works of philosophy and theology throughout his life, his most influential work is the *Summa Theologica.* Aquinas's economic thought is inseparable from his understanding of natural law. In his view, natural law is an ethic derived from observing the fundamental norms of human nature. These norms can be understood as the will of God for creation. An unlawful act is that which perverts God's design for a particular part of His creation. Economic transactions, according to Aquinas, should be considered within this framework, since they occur as human attempts to obtain materials provided by nature to achieve certain ends.

Private property is a desirable economic institution because it complements man's internal desire for order. "Hence the ownership of possessions is not contrary to the natural law," Aquinas writes in the Summa Theologica, "but an addition thereto devised by human reason." The state, however, has the authority to maintain a legal framework for commercial life, such as enforcing rules prohibiting theft, force, and fraud. In this way, civil law is a reflection of the natural law. Further, Aquinas believed that private ownership of property is the best guarantee of a peaceful and orderly society, for it provides maximum incentive for the responsible stewardship of property. Aquinas helped relax the traditionally negative view of mercantile trade that figured prominently in, for example, Patristic thought. For Aquinas, trade itself is not evil; rather, its moral worth depends on the motive and conduct of the trader. In addition, the risk associated with bringing goods from where they are abundant to where they are scarce justifies mercantile profit. The merchant, however, must direct his profits toward virtuous ends.

On Exchange and Value: To Aquinas the issue of exchange was of significant importance in barter situations. It was his belief that the measure of exchange between heterogeneous goods indicated that some value must be placed on each good and that an ethical dilemma would arise. The only just exchange would be that of two identical goods for the same quantity, but this would be redundant; therefore, some value would have to be established for the goods in order for the exchange to be just to both buyer and seller (or in his case traders, being that little to no market existed for surplus commodities). This posed, for him, the ethical aspect of prices, raising issues of equity and justice".

On Usury: Usury is defined today as the charging of excessive interest on a loan. During the reign of the Scholastics usury applied to any charging of interest. During the Medieval times a loan was usually of a good, and was hardly if ever intended as a method of producing any wealth. The Second Lateran Council (1139) banned usury using direct scriptural reference as their cause. Both Aristotle and Aquinas argued that the taking of any interest on loans was unjust and in conflict with natural law.

On Good and Justice: Central to the theories of the Scholastics is the notion that man is directed through his interactions with his society by free will, and autonomy. This moral philosophy has been introduced into economics by Aquinas, Adam Smith, J.S. Mill, and Karl Marx to name a few. "It is a matter of common experience that our conduct is motivated by different aims: riches, honor, material pleasure, social positions, etc.". It is the reconciliation of these aims with what is best for us and our community that reveals the true good. Servitude or slavery for example is good for the man exercising authority over another and may be good for the servant for lack of any other means of existence. The private utility of another, however, is unjust and denies the respect and will of the other. "As much as Karl Marx, St. Thomas is cognizant of the humiliation inflicted on man by what Marx calls the alienation of work for the profit of another, and what St. Thomas called more simply servitude".

19.1.1. Scholasticism

Scholasticism (The Schoolmen)refers to the school of economic thought that developed in Europe during the medieval period (500-1500). Scholastic thinkers are known for their moral and philosophical approach to the study of exchange, value, and ownership within the context of the time period. Of the scholastics, St. Thomas Aquinas is widely credited for his original, although sometimes ambiguous contributions to the early discussions of value, price, private property, and usury (or interest). Until the arrival of Mercantilism in the 14th century the Scholastics (or Schoolmen as they are commonly referred to today) were at the forefront of the foundations of establishing economic theory within the framework of philosophy. Probably the most influential economic thinker of the Scholastic period was a Sicilian-born Roman Catholic by the name of Thomas Aquinas. Scholasticism evolved amongst a societal structure known as feudalism. The feudal society of Medieval Europe was one in which all authority was derived from God by the church, which was headed by the Pope. The feudal system was one in which the king of a land or region delegated power, responsibility, and land grants to his royal subordinates (nobles, barons, lords, etc.). These barons would in turn sub-let land to landlords, with the understanding that the baron had full control of the land, established his own laws and taxes, and had the right to call all to serve under the crown. It was then the lowly serfs whose duty it was to tend the land and provide free labour, food, and service whenever itwas demanded. The relationship between lord and serf was one dictated by custom, tradition and authority. It was this class relationship that caused religious theologians of the time to examine what moral and ethical implications were addressed when two or more parties entered into an exchange, or contract.

19.1.2. Mercantilism

Mercantilism is economic nationalism for the purpose of building a wealthy and powerful state.Adam Smith coined the term "mercantile system" to describe the system of political economy that sought to enrich the country by restraining imports and encouraging exports. This system dominated Western European economic thought and policies from the sixteenth to the late eighteenth centuries. The goal of these policies was, supposedly, to achieve a "favorable" balance of trade that would bring gold and silver into the country and also to maintain domestic employment. In contrast to the agricultural system of the physiocrats or the laissez-faire of the nineteenth and early twentieth centuries, the mercantile system served the interests of merchants and producers such as the British East India Company, whose activities were protected or encouraged by the state. The most important economic rationale for mercantilism in the sixteenth century was the consolidation of the regional power centre of the feudal era by large, competitive nation-states. Other contributing factors were the establishment of colonies outside Europe; the growth of European commerce and industry relative to agriculture; the increase in the volume and breadth of trade; and the increase in the use of metallic monetary systems, particularly gold and silver, relative to barter transactions.

During the mercantilist period, military conflict between nation-states was both more frequent and more extensive than at any other time in history. The armies and navies of the main protagonists were no longer temporary forces raised to address a specific threat or objective, but were full-time professional forces. Each government's primary economic objective was to command a sufficient quantity of hard currency to support a military that would deter attacks by other countries and aid its own territorial expansion. Most of the mercantilist policies were the outgrowth of the relationship between the governments of the nation-states and their mercantile classes. In exchange for paying levies and taxes to support the armies of the nation-states, the mercantile classes induced governments to enact policies that would protect their business interests against foreign competition.

Alexander Gray observes that mercantilism is a misleading and deceitful word. Different writers have defined mercantilism differently. According to Lekachman "mercantilism was a battle against hampering medieval thought and practice". It was revolt against medievalism resolve to reconstruct economic life to a more rational scheme. To Edmund Whittaker mercantilism was the economic counterpart of political nationalism". Heiman described it as the ideological justification of Commercial Capitalism. Thus mercantilist writers were essentially practical businessmen, merchants and administrators in various European countries like England, France, Italy, Germany, Scotland, Spain etc. They left behind numerous works regarding contemporary national economic problems. They do not form a school of economists. So the ideas and policies which dominated the economic scene of England 'and a part of Europe between the close of the 16th century and the middle of the 18th century can rightly be called as mercantilism. Mercantilist writers put emphasis on foreign trade as a means of accumulating treasure and building a strong nation. The growth of Mercantilism was the result of combination of factors cultural, religious, political and economic and it shall be desirable to examine these causes in some details. In the beginning of the sixteenth century Europe witnessed great religious and intellectual awakening due to Reformation and Protestantism. These two movements associated with the names of Erasmus and Martin Luther respectively which gave a great fillip to the ideas of individualism and personal freedom and went a long way in developing the concepts of property and contract rights which in turn led to the growth of commerce and free exchange.

Before the emergence of these movements, the Pope enjoyed a predominant position in religion and could also interfere in the worldly matters. With the rise of Protestantism the monetary aspect of life was emphasized and a bid was made to confine the authority of the Pope to the religious matters alone and prevent his interference in the economic and political matters. Even the international position of the church was challenged by setting up national churches. For example, in England Henry VIII seized the church property and established the Church of England and himself became its spiritual head. Renaissance played even more significant role and highlighted the element of humanism. It challenged the medieval theologian concept that happiness in heaven should be preferred over worldly happiness, and asserted that happiness on this earth was to be preferred over the promised pleasures of the other world.In other words, it emphasized the materialistic basics of the human happiness. Once the principles of humanism and individualism were accepted, a large number of writers, artists, philosophers emphasized the economic basis of the society in their works and shook the foundations of the edifice of Church Theology.

In the economic sphere the decline of feudalism greatly contributed to the growth of mercantilism. The feudal system was characterized by economic self-sufficiency, agricultural production and absence of exchange economy. The agriculturists were required to work free of charge on the fields of the lords for a stipulated period. They were also required to work as soldiers for the lords in times of war. As there were no organized industries and even commercial crops were not in much demand, these agriculturists worked for local self-sufficiency in food grains. In the absence of organized markets the manufacture was undertaken chiefly to meet the local requirements. This resulted in the growth of an independent domestic economy based on local self-sufficiency. Above all, there was no effective state organization. In the cities and towns the guilds and municipalities tried to regulate the trade between different localities. However, with the expansion of commerce divergent individual trading interests came to the fore. Almost all of them looked for a strong central authority to protect them against their rivals. In the absence of a national government this was not possible and the relationship was decidedly a weak link. The growth of commerce and development of domestic economy gave rise to the problem of labour and distribution. But probably the most important factor which stimulated the development of mercantilism was the emergence of the exchange economy. This led to development of international trade, which in turn encouraged large scale production. For a fuller utilization of the available economic resources it was felt that the economic life should be regulated. The urge for new marks led to the discovery of new islands and countries and the development of colonialism. In short, we can say that mercantilism was stimulated by factors like decline of feudalism, lack of state organization, rise of free labour classes, competition and development of exchange economy. Many factors are responsible for the rise of mercantilism. So to understand the basic tenets of mercantilists, it is necessary to study first the conditions like social, economic political cultural and scientific and the changes that were taking place during the 13th and 14th centuries in Europe.

The important factors are given below:

(1) **Economic Factor:** Towards the end of the 15th century, economic changes were taking place in European countries. The important economic change was the decline of feudalism and its gradual replacement by Commercial Capitalism. The domestic self-sufficiency was giving way to the development of a system of exchange. The market economy was slowly emerging, replacing the feudalistic self-sufficient local economy. Agriculture was being replaced by trade and commerce. Monetary transactions were rapidly expanding. The feudal society was rapidly breaking up and commercial system was gradually evolving. This paved the way for the rise of a commercial system using money.

(2) **Political Factors:** The major political change was the transition from feudalism to nationalism. Building up a strong nation state was in the forefront. In many European countries feudalism was superseded by the formation of nation states. These countries required a strong Government. Machiavelli and Jean Bodis postulated the need for unified national state under a strong king- The merchants and trading community were looking forward for a strong centre to protect them from rivals and to nourish their trading interests.

(3) **Religious Factors:** The ideas of Reformation and protectionism spread individualism and personal freedom. The Reformation movement revolted against the supremacy of Roman Catholic Church and the authority of me Pope. These religious changes helped a lot in developing property rights and contract rights which were essential for the growth of commerce. The new Protestant religion allowed the acquisition of material goods

and property. The new religion was in favour of complete liberty and freedom of the individuals. The Protestant religion became very strong in course of time and it was embraced by the merchants.

(4) **Cultural Factors:** Medieval theology laid emphasis on heavenly happiness. It preached the people to detach from material wealth. Against this two important cultural changes arose in the society. They are renaissance and humanism. These cultural changes gave emphasis to creative human activities, acquisition of wealth, and trade and commerce. This new ideology was able to uphold the spirit of the traders and merchants to accumulate wealth. In fact mercantilism was a reaction against the moral and idealistic attitude of the medieval period.

(5) **Discoveries and Inventions:** Various discoveries and inventions played an important role in the promotion of trade and commerce and helped the development of mercantilist thought. *e.g.* invention of Mariner's compass, Columbus Discovery of America, Discovery of Gold and silver mines in the new world, the invention of printing press, new sea route to India etc. helped a lot for the expansion of foreign trade and development of foreign markets.

All the above factors helped the growth of mercantilism which was a new adventure in the direction of the commercial merchant capitalism in the European countries.

19.1.3. Important Economic Theory of Mercantilism

Whereas the economic literature of scholasticism was written by medieval churchmen, the economic theory of mercantilism was the work of secular people, mostly merchant businessmen, who were privately engaged in selling and buying goods. The literature they produced focused on questions of economic policy and was usually related to a particular interest the merchant and writer (in one person) was trying to promote. For this reason, there was often considerable scepticism regarding the analytical merits of particular arguments and the validity of their conclusions. Few authors could claim to be sufficiently detached from their private issues and offer objective economic analysis. However, throughout the mercantilism, both the quantity (there were over 2000 economic works published in 16th and 17th century) and quality of economic literature grew. The mercantilist literature from 1650 to 1750 was of distinctly higher quality, these writers created or touched on nearly all analytical concept on which Adam Smith based his Wealth of Nations, which was published in 1776. The age of mercantilism has been characterized as one in which every person was his own economist. Since the various writers between 1500 and 1750 held very diverse views, it is difficult to generalize about the resulting literature. Furthermore, each writer tended to concentrate on one topic, and no single writer was able to synthesize these contributions impressively enough to influence the subsequent development of economic theory.

Secondly, mercantilism can best be understood as an intellectual reaction to the problems of the times. In this period of the decline of feudalism and the rise of the nation-states, the mercantilists tried to determine the best policies for promoting the power and wealth of the nation, the policies that would best consolidate and increase the power and prosperity of the developing economies. What is especially important here is the mercantilistic assumption that the total wealth of the world was fixed and constant. These writers applied the assumption to trade between nations, concluding that any increase in the wealth and economic power of one nation occurred at the expense of other nations (the rest of the world). Thus, the mercantilists emphasized international trade as a mean of increasing the wealth and power of a nation. Using some modern game-theoretic language, we may say, that they perceived economic activity and international trade in particular as a zero-sum game, that is a game, where it is impossible for both players to win (In a two-person zero-sum game, the payoff to one player is the negative of that going to the other player). So according to mercantilists, it is impossible to increase a global wealth of the world in effect of international trade. It is a very sad assumption, and modern economists do not share it. The goal of economic activity, according to most mercantilists, was production, not consumption, as classical economists would later have it. They advocated increasing the nation's wealth by simultaneously encouraging production, increasing exports and holding down domestic consumption. Thus, in practice, the wealth of nation rested on the poverty of the many members of society. One again, they advocated high level of production, high level of export and low domestic consumption. In addition, they proposed low wages in order to give the domestic economy competitive advantages in international trade. Most of the mercantilists held that wages (of the workers) should be set on the subsistence level, allowing workers to preserve their lives, but not to consume more than it is required to continue their lives. Higher wages would cause labourers to limit their work supply and national output; national wealth would fall, according to mercantilists. Thus, when the goal of economic activity is defined in terms of national output and not in terms of national consumption, poverty for the masses benefits the nation. This is another sad consequence of mercantilist economic theory.

Third general point about mercantilism is their insistence on the notion of balance of trade. Balance of trade figures, also called net exports, are the sum of the money gained by a given economy selling exports, minus the cost of buying imports. A positive balance of trade is known as a trade surplus and consists of exporting more (in financial terms) than one imports. A negative balance of trade is known as a trade deficit and consists of importing more than one export. As we know today, neither positive nor negative balance of trade is necessarily dangerous in modern economies, although large trade surpluses or trade deficits may sometimes be a sign of other economic problems. According to mercantilists a country should increase exports and discourage imports by means of tariffs, quotas, subsidies, taxes and the like in order to achieve a so-called favourable or positive balance of trade. Production should be stimulated by government interference in the domestic economy and by the regulations of foreign trade. Protective duties should be placed on manufactured goods from abroad; and the state should encourage the import of cheap raw materials to be used in manufacturing goods for export.

Fourth, the mercantilists argued for a positive balance of trade. Many early mercantilists defined the wealth of nation not in terms of nation's production or consumption, but in terms of its holdings of precious metals such as gold or silver. They argued for positive balance of trade because it would lead to a flow of precious metals into the domestic economy to settle the trade balance. Remember here that later or most eminent mercantilists equated the wealth of nation with the overall production of the nation. While for early mercantilists the wealth of nation consisted of the precious metals accumulated in the domestic economy. The first mercantilists argued that a positive balance of trade should be struck with each nation.

A number of subsequent writers however argued that only the overall balance of trade with all nations was significant. Thus, England might have a trade deficit with India, but because it could import from India cheap raw materials that could be used to manufacture goods in England for export, it might well have a positive overall trade balance with all nations.

A related issue concerned the exports of precious metals or bullion (that is gold or silver in bars). Early mercantilists recommended that the export of bullion be strictly prohibited. Later writers suggested that exporting bullion might lead to an improvement in overall trade balances if the bullion were used to purchase raw materials for export goods.

Fifth, about mercantilist theory concerns money. The early mercantilists equated the wealth of nations with the stock of precious metals internally held in the country. They were very impressed with the significance of the tremendous flow of precious metals into Europe, particularly into Spain, from the New World, from America. It is therefore not surprising that they became to identify the wealth of nations with gold and silver. However later mercantilists subscribed to a more sophisticated view and identified the wealth of nation with the nation's overall production. They were able to develop useful analytical insights into the role of money in an economy. A central feature of this late mercantilist literature is the conviction that monetary factors, money supply, rather than real factors (such as quantity of labour, capital goods and the like), are the chief determinants of economic activity. They maintained that an adequate supply of money is particularly essential to the growth of trade, both domestic and international. Changes in the quantity of money, they believed, generate changes in the level of real output. Therefore, in this view a positive balance of trade, which would effect in a flow of money into the domestic economy, would increase the production, the real output and therefore contribute to the increasing wealth of nation. Classical economists and Adam Smith radically rejected this view, that monetary factors are main determinants of economic activity and economic growth in the second half of 18th century. Classical economists held that the level of economic activity and the rate of growth depend upon a number of real factors: the quantity of labour, natural resources, capital goods and the institutional structure of the society. Any changes in the quantity of money according to classical economists would not influence the level of neither output nor growth, but only the general level of prices.

Mercantilists held that money supply (not any non-monetary, real factor) was the main factor contributing to the wealth of nation. So much on the assumptions of the mercantilist economic doctrine, and we can move to the discussion of its contribution to the economic theory – theoretical contributions of mercantilists.

Probably the most significant accomplishment of the later mercantilists was the explicit recognition of the possibility of analysing the economy. As we remember, the ancient and medieval thinkers were mostly engaged in moral, ethical analysis of the economy. It is just in mercantilist writings, that economy becomes an object of a purely scientific, not moral analysis. They started to think about an economy as a subject of science, they realized that the laws of economy could be discovered by the same methods that revealed the laws of physics and other sciences. This was extremely important step toward subsequent developments in economic theory. Many mercantilists saw an economy as a mechanical system and believed that if one understood the economic laws governing the economy, one could control the economy. Wisely proposed legislation could, in this view; positively influence the course of economic events and economic analysis would indicate what forms of government intervention would help the economy. They however realized that government intervention must not be in contradiction with some basic economic truths such as the law of supply and demand. Some of them correctly deduced, for example, that price ceilings set below the equilibrium prices lead to excess demand and shortages. In addition, the later mercantilists frequently applied the concepts of economic man and the profit motive in stimulating economic activity – they said, that government cannot change the basic nature of human beings, and especially their egoistic drives. Therefore, politicians have to take these factors as given (that humans are egoistic and driven by profit motive) and try to create a set of laws and institutions that will channel these drives to increase the power and prosperity of the nation.

Many of the later mercantilists became aware of the serious analytical errors of their predecessors (that is early mercantilists). They recognized, for example, that stock of gold and silver is not a measure of the wealth of a nation, that it was not possible for all nations to have a positive balance of trade, and also that no one country could maintain a positive balance of trade over the long run. What's more, some of them realized that trade can be mutually beneficial to nations and that advantages will occur in those countries that practice the division of labour in production. An increasing number of writers recommended a reduction in the amount of government intervention, anticipating the prescriptions of Adam Smith and other classical economists. We have to remember however, that none of these writers was able to present an integrated view of the operation of a market economy –the manner in which prices are formed and scarce resources are allocated. This was eventually achieved by Adam Smith and classical economists in the second half of 18th century and in 19th century. The popular explanation of this failure of mercantilists to produce a systematic account of a market economy is that the mercantilists believed that there was a basic conflict between private and public interests or between private and public welfare. Therefore, they considered it necessary for government to channel private self-interest into public benefits. Classical economists, on the other hand, found a basic harmony in society between private, egoistic drives of individuals and social welfare. Even the later mercantilists who advocated market policies lacked sufficient insights into the operation of market to make an adequate argument linking private self-interest and social welfare. Still, the writings of the later mercantilists were used by Smith to develop his analysis, and this is another significant contribution of the mercantilists.

Groups in 1914 into a larger economy for exploitation for the benefit of British industrial classes. Under colonial rule, Nigeria remained an agricultural country, exporting raw materials to Britain and importing from it finished goods. Therein lay the origins of the dependence of Nigerian economy on commodity markets of the industrialized Western world for its foreign exchange. While the industrialization of the country was discouraged, rudimentary foundations for a modern Nigerian economy, however, were laid. Colonial economic policies shaped future independent Nigeria's economy, particularly in marketing, labour supply, and investment. The process of colonial rule and formal economic exploitation ended in 1960 but left Nigeria a relatively strong but undiversified economy. Thereafter, Nigerians were poised to remedy this defect and to build a self-sustaining Nigerian economy comprising agricultural, industrial, and service sectors.

20
Economic Ideas of Iben Khaldun

20.1. Economic Ideas of Iben Khaldun

Ibn khaldun is a fourteenth century Muslim thinker. Ibn Khaldun was born in Tunis on 27th May 1332 AD. He acquired his great education in Islam, logic, philosophy, law, grammar and poetry, all of which were to contribute towards the moulding of a great statesman-like ability within him. Ibn Khaldun has dealt with economics, sociology, political science and other subjects in order to understand the behaviour of man and his history. The contributions of Ibn Khaldun to the development of economic thought have gone largely unnoticed in the academic realm. He has written on many subjects, including on the rise and fall of nations in his Muqaddimah: An Introduction to History. His writings on economics, economic surplus and economic oriented policies are as relevant today as they were during his time. His emphasis on less government expenditure for mercenary army has been heard by many developed countries which are in the process of implementing his policy prescriptions in order to increase economic surplus by shifting resources to education and human development. He opposed taxation and tariffs that discouraged trade and production. Ibn Khaldun opposed state involvement in trade and production activities. He thought the bureaucrats cannot understand commercial activities and they do not have the same motivations astradesmen. He predicts relative fall of economic surplus and the decline of the countries in which state involves in trade and production. He sees a large army as impediment to the expansion of trade, production and economic surplus.

Ibn Khaldun is the first person who has systematically analyzed the functioning of an economy, the importance of technology, specialization and foreign trade in economic surplus and the role of government and its stabilization policies to increase output and employment. Ibn Khaldun, moreover, dealt with the problem of optimum taxation, minimum government services, incentives, institutional framework, law and order, expectations, production, and the theory of value. Ibn Khaldun, again is the first economist with economic surplus at hand, who has given a biological interpretation of the rise and fall of the nations. His coherent general economic theory constitutes the framework for his history. No one in the history of economic thought has established such a coherent general economic theory to explain and predict the rise and the fall of the civilizations, nations and the empires as Ibn Khaldun has formulated in his Muqaddimah: An Introduction to History. His theory has the empirical and theoretical power not only to explain the consequences of government policies on production and trade, investment and specialization, but to predict the very survival of the state. On specialization and economic surplus: Ibn Khaldun had indicated the fact that specialization is the major source of economic surplus, almost three centuries before Adam Smith. For Ibn Khaldun, when there is an environment conducive for specialization, the entrepreneur is encouraged to commit himself for further trade and production. Indeed, specialization would occur in a place in which a person is able to get the benefit of his efforts. Given law and order, for him, specialization is a function of population, trade, production and minimum taxation. The concepts of mass production, learning by doing and theconcept of on the job training have been exposed by Ibn Khaldun in the above statement so clearly which needs no further clarification. However, it is important to indicate that these very concepts, Ibn Khaldun had dealt in his writings, had become the subjects of articles in Economic Literature in the late 50's. For Ibn Khaldun, specialization meant the coordination of different functions of factors of production where, "what is obtained through the cooperation of a group, of human beings satisfies the need of a number many times greater (than themselves)".

On supply and demand: Ibn Khaldun, again centuries ahead of his time, postulated that prices of goods and services are determined by supply and demand. When a good is scarce and in demand, its price is high. The merchant will buy the goods "where they are cheap" and plentiful and "selling them at a high price" where they are scarce and in demand. Naturally, when a good is plentiful, its price is low: "the inhabitants of a city have more food than they need. Consequently, the price of food is low, as a rule, except when misfortunes occur due to celestial conditions that may affect (the supply of) food" Moreover, he had the concept of long run cost of production in the Marshallian sense. On monetary policy: Ibn Khaldun defends a stable monetary policy. He is against the policies of the authorities to play with the value of currency. He fears that the authorities may be tempted to debauch with the value of money in order to build palaces and finance mercenary armies. This process will cause inflation. Population will lose confidence in currency. And these developments are considered to be unjust. As a supreme policy for the society, the protection of purchasing power of money has to be implemented as a matter of justice. To do that, he proposed an independent monetary agency under the authority of Chief Justice, a "God-fearing man" to prevent the rulers "fearlessly" from tempting with the value of money and debauching the currency. For him, as far as monetary policy is concerned, a stable monetary policy aiming the protection of purchasing power of money is a must as a matter of justice. The population has to be protected from unjust policies of the rulers when they debauch the currency. A stable and sound currency increases the confidence of people in currency, trade and production. For him, what is needed for the society is less government expenditure on palaces and bureaucracy, less expenditure on mercenary army, less taxation and a stable currency for trade and production.

On fixed prices: Ibn Khaldun was not only against state involvement in commercial and agricultural activities, he was also against government's involvement in fixing the prices of goods and services. When the government employs force "by buying things up at a cheapest possible price", the ruler "will be able to force the seller to lower his price" and "forces the merchants or farmers who deal in these particular products to buy from him". The rulers "undertake to buy agricultural products and goods from their owners who come to them, at prices fixed by them as they see fit. Then, they resell these things to the subjects under their control, at the proper times, at prices fixed by them. Ibn Khaldun' Economic Prescriptions for A Civilized Society as given in 'Moqaddimah: Introduction to History: The Rise and Fall of Nations (1350)':

Given political stability and solidarity (asabiyah), for the rise of the nations, there must be:

- ✰ A firm of establishment of private property rights and freedom of enterprise
- ✰ Rule of law and the reliability of judicial system for the establishment of justice
- ✰ The security of peace and the security of trade routes
- ✰ Lower and less taxation in order to increase employment, production and revenues
- ✰ Less bureaucracy and much smaller efficient army
- ✰ No government involvement in trade, production and commercial affairs
- ✰ No fixation of prices by the government
- ✰ A rule that does not give monopoly power to anyone in the market
- ✰ Stable monetary policy and independent monetary authority that does not play with the value of money
- ✰ A larger population and a larger market for greater specialization
- ✰ A creative education system for independent thinking and behavior
- ✰ The collective responsibility and internal feeling for the setting up of a just system to encourage good deeds and prevent vice.

20.2. Criticisms of Marx's Theory

Following are criticisms that are commonly made:

- Too much emphasis is given to the economic factor in explaining social order and change. Culture seemed to be explained solely as derived from the economic 'substructure'. However it has a degree of 'autonomy'; for example it is difficult to explain the advent of gay liberation in terms of productive or economic relations.
- Even if you get rid of capitalism you might still have enormous problems of conflict and domination in society. State bureaucracies as well as capitalists can dominate – ask the Russians.
- Marx's theory of history is contradicted by the fact that industrialized countries have not moved closer to revolution. The recent revolutions have been in peasant societies, such as China. Capitalist societies seem to have become more secure from threat of revolution throughout the 20th century.
- Many would say there are no laws of history and that Marx was mistaken in thinking he had discovered the laws of history, and in thinking that his theory was scientific.
- Anarchists say Marxists fail to grasp the unacceptable dangers in their readiness to take an authoritarian-centralist approach. Marxists are willing to use the authoritarian state to run society after the revolution and to be ruthless in this. This is extremely dangerous; those in control can't be trusted and are very likely to become an entrenched dictatorship (*e.g.*, Stalinism).
- Many if not all Anarchists would also reject Marx's theory of how capitalism can or will be replaced, which involves confronting capitalism, class conflict, seizing the state and taking power from the capitalist class, and destroying capitalism, a process which will probably involve violence. However some anarchists believe the change could come via increasing awareness and disenchantment, the building of alternative communities based on anti-capitalist principles, and thus an increase in the numbers who want to abandon capitalism especially given that its coming difficulties will probably increasingly reveal its inability to provide for all.
- Marx (and most Marxists today) failed to take ecological sustainability into account. They are strong believers in industrial development and 'progress', rising material 'living standards' and economic growth. They think that capitalism is responsible for all problems and that when it has been eliminated we can release the previously restricted power of industry and eliminate waste to enrich everyone. In other words, Marxism has no concept of 'limits to growth'. Affluence and economic growth are regarded as desirable and possible. 'Dark green' critics insist that a good, post-capitalist society cannot be a growth society, and it cannot have high per capita levels of resource consumption. Getting rid of capitalism is not enough; there is even bigger problem, set by the commitment to industrialism, growth and affluence. Marx could not have known that a time would come when we would run into a problem of over-consumption.

From the perspective of 'The Simpler Way' a high quality of life for all is achievable without high material 'living standards' or much modern technology, let alone industrialization and IT etc. We have no doubt that Marx was mistaken about scarcity, the stinginess of nature and the impossibility of human emancipation before technical advance delivered material abundance. His concept of development was really the same as capitalist modernization. That is capital is crucial for development. He dismissed peasant ways and Marxists are not sympathetic to the notion of 'appropriate development' conceived mainly in terms of 'subsistence' and low/intermediate technology and cooperative ways focused on local economic self-sufficiency.a Gandhian way. In other words, advocates of The Simpler Way claim Marx was quite mistaken in thinking that socialism would not be possible without modern technology, industrialization and material affluence. Achieving a good society does not require elaborate technology nor abundance. It depends on whether or not the right values are held. There have been societies, and there are societies today in which people live well with very humble material lifestyles and without modern technology.

- ✰ Marxist ideas on how to change society are also strongly criticized by the Anarchists. Marxists thought capitalism must be fought and overthrown through violent revolution, because the capitalist classes will never voluntary give up any of its privileges. There must be leadership by a vanguard party prepared to be ruthless and to use violence, and which will rule in an authoritarian way after the revolution. Eventually when a person have developed the right ideas and values the state can dissolve and there will be a communist society.

20.3. Nigeria Experience: Overview of Economy

As of 2001, the most conspicuous fact about Nigeria's economy is that the corruption and mismanagement of its post-colonial governments has prevented the channeling of the country's abundant natural and human resources–especially its wealth in crude oil–into lasting improvements in infrastructure and the construction of a sound base for self-sustaining economic development. Thus, despite its abundant resources, Nigeria is poorer today than it was at independence in 1960. Still one of the less developed and poorer countries of the world, it has the potential to become a major economic power if the leaders resolve to learn from past mistakes and to harness the country's rich natural and human resources for a productive and sustained effort to promote economic development. Before the country was colonized by Britain, during the second half of the 19th century, the various nationality groups that currently make up Nigeria were largely an agricultural people. They were food self-sufficient and produced a variety of commodities that were exported overseas. British colonial administrators amalgamated (joined together) the nationality.

From independence in 1960, the state took up the direction and planning of economic growth and development. Education was progressively expanded at all levels to reduce the rate of illiteracy and to provide the requisite skills and labour force for development. Infrastructure of roads and communication networks were constructed far beyond what was inherited from colonial rule. Hydroelectric dams were built to generate electricity. Secondary industries and automobile assembly plants were established to create more employment opportunities. Because of the paucity (small number) of indigenous (native or local) private capital, these activities were undertaken and financed by the government, often with foreign assistance from such countries as Britain and the United States. Foreign oil companies, such as Shell-BP, Exxon-Mobil, Chevron, Agip, and Texaco, operate in partnership with the government in the oil sector, the mainstay of Nigeria's economy. The capital-intensive oil sector provides 95 percent of Nigeria's foreign exchange earnings and about 65 percent of its budgetary revenues.

Because the established, government-owned industries and businesses were often inefficient and corrupt, productivity was low at best. In particular, mismanagement and corruption were endemic (characteristic of) in the successive governments and throughout the nation. However, the gravest problem was caused by the government's decision to stress the industrial sector above all others. Caught in a web of competing demands for scarce resources, the officials took the path of rapid, large-scale industrialization at the expense of the agricultural sector, as well as light manufacturing. They directed the bulk of investment capital towards the promotion of what Western advisers captioned "industrial take off." This decision to abandon the known–agriculture–for the unknown–rapid large-scale industrialization–was a fundamental error. The capital and the skill needed for rapid, large-scale industrialization were not sufficiently available. Thus, an unskilled workforce and insufficient funds severely handicapped the industrial sector. Also, Nigeria's neglect of the agricultural sector aggravated already problematic food shortages. Nigeria had raised enough food to meet domestic needs during its colonial period and in the decade following independence. However, it experienced food shortages in the 1970s and 1980s, which necessitated the importation of food from foreign countries. Among the imports were palm oil (from Malaysia), of which Nigeria had been the world's largest producer and exporter, and rice (from the United States) which was considered less nutritious than Nigerian brown rice. Once Africa's largest poultry producer, Nigeria lost that status because of inefficient corn production and a ban on the importation of corn. Furthermore, it is no longer a major exporter of cocoa, peanuts, and rubber.

Several forces compounded the problems of the agricultural sector. The migration of labour from the rural areas to the urban centers reduced the traditional agricultural labour force. Ecological constraints such as poor soil, erosion, drought, and the absence of agricultural research added to the problem. Other constraints on agricultural production include the use of antiquated technology due to a lack of capital, the low status given to agriculture in the education of the youth, inefficient marketing, an inadequate transportation infrastructure, lack of refrigeration, trade restrictions, under-investment due to unavailability of credit, low prices, and unstable pricing policies which resulted in farmers literally subsidizing urban dwellers and other sectors of the economy. In addition to these handicaps, import constraints limit the availability of many agricultural and food-processing plants. In general, land tenure discourages long-term investment in technology and modern production techniques.

The problem of food shortages and imports was addressed in the late 1970s and early 1980s. In the late 1970s the military government of Olusegun Obasanjo embarked upon "Operation Feed the Nation." His civilian successor, President Shehu Shagari, continued the program as the "Green Revolution." Both programs encouraged Nigerians to grow more food, and urged unemployed urban dwellers to return to the rural areas to grow food crops. The government provided farmers with fertilizers and loans from the World Bank. The food situation has stabilized, although Nigeria still imports food. A related problem which has not been completely resolved is the pollution of water in the Delta region and Ogoniland by oil companies. Water pollution disrupts farming efforts and has been a source of friction between farmers on one side and the national government and the oil companies on the other.

The oil boom which Nigeria experienced in the 1970s helped the nation to recover rapidly from its civil war and at the same time gave great impetus to the government's program of rapid industrialization. Many manufacturing industries sprang up and the economy experienced a rapid growth of about 8 percent per year that made Nigeria, by 1980, the largest economy in Africa. The growth, however, was not sustained. The new oil wealth did little to reverse widespread poverty and the collapse of even basic infrastructure and social services. The iron and steel industry, started with the help of the Soviet Union, still has not achieved a satisfactory level of production. The oil boom also provoked a shortage of labour in the agricultural sector as members of the rural workforce migrated to jobs in the urban construction boom and a growing informal sector. When the price of crude oil fell and corruption and mismanagement still prevailed at all levels, the economy became severely depressed. The urban unemployment rate rose to 28 percent in 1992, and crime also increased as 31.4 percent of the population lived below the poverty line.

Nigeria's debts mounted as administrators engaged in external borrowing and subsidized food and rice imports and gasoline prices. In the 1980s, economic realities forced Ibrahim Babangida's military regime to negotiate a loan with the World Bank and to reschedule Nigeria's external debts. His regime undertook an economic structural adjustment program (ESAP) to reduce Nigeria's dependence on oil and to create a basis for sustainable non-inflationary growth. However, external borrowing to shore up the economy created more problems than it alleviated. Much of the borrowed money never reached Nigeria. The portion that reached the country often went towards abandoned or nonperforming public sector projects. External loans escalated Nigeria's debts to US$30 billion during the Babangida regime and consumed external earnings in debt servicing. Similarly, the ESAP prescribed by the International Monetary Fund (IMF) failed to advance the economy, and aggravated the problems of inflation and unemployment. It caused a reduction of state spending on education and health care. Continuing political instability due to Babangida's annulment of the presidential election results in June 1993 and the subsequent authoritarian rule of Sani Abacha (1993 to 1998) made the general economic situation worse. The gross corruption by the Abacha regime and its violations of people's fundamental rights turned Nigeria into an international pariah for 6 years, and thus discouraged foreign investment in the economy. Many industries and manufacturing companies could not obtain raw materials and closed down. Others operated under severe handicaps, including rampant power outages and refined petroleum scarcity. Not enough had been done in the years of plenty to diversify the economy or to sustain the development. Military coups (military overthrow of civilian governments) and political instability worsened the situation. There was considerable optimism in May 1999 when Oluseguan Obasanjo became Nigeria's civilian president. Many hoped that he would lift Nigeria from the verge of economic bankruptcy. One of Obasanjo's objectives to that purpose was to secure debt relief from Nigeria's foreign creditors. However, these creditors insisted that Nigeria's wealth of untapped resources provided the means for the country to pay off its debts, and refused to cancel its debts of US$30 billion.

In spite of some opposition, Obasanjo embarked upon a program of privatizing some parastatals in order to reduce corruption, promote efficiency, and raise productivity. He introduced an anti-corruption bill which passed through the legislature, and recovered some of the revenues that had been stolen from the country and deposited in Western banks. The inflation rate, which was estimated at 12.5 percent at the start of his administration, was estimated at 6.6 percent in 2000. Significant exports of liquified natural gas started in 1999, and increased crude oil prices in 2000 provided his administration with additional revenues. So far, however, he has been unable to bring about economic recovery. Industrial capacity utilization appears to have diminished. Worse still, infrastructural facilities, including the National Electric Power Authority (NEPA), continue to be in a state of disrepair. Expected massive inflow of foreign investment, on which the government had hinged its economic revival program, failed to materialize. This is in part because of the high cost of doing business in Nigeria and a lack of transparency in economic decision-making in the country. In addition to these realities, the unemployment situation in the country remains unchanged months after the restoration of civilian government. In fact, it has worsened among university graduates and ranged from 30 to 40 percent in 2000. Political uncertainties due to ethnic and religious conflicts between Muslims and Christians, and constant feuding between the president and the legislators aggravate the economic climate. Widespread armed robbery and a crime syndicate known locally as 419 prey on foreign nationals, further hindering foreign investment and tourism. The country's economy needs the collective efforts of the president and the National Assembly as well as more definite measures to address its ills in order to foster its recovery and growth. Currently, funds available to the government are insufficient to meet the needs of all sectors of the economy at once. External investors can contribute through long-term investment and **joint ventures** in Nigeria's large national market. Crude oil, the price of which rose sharply recently, remains a very considerable asset. Properly managed, it could provide a solid platform for more sustained Nigerian development and prosperity in the 21st century and beyond.

References

Ackerman, Frank & Lisa Heinzerling, 2004. *Priceless. On Knowing the Price of Everything and the Value of Nothing*. The New Press, New York.

Aizenman, J., and N. P. Marion, 2003. "The High Demand for International Reserves in the Far East: What's Going On?" *Journal of the Japanese and International Economies*, Vol. 17, No. 3, pp. 370–400.

Aizenman, J., and N. P. Marion, 2004, "International Reserves Holdings with Sovereign Risk and Costly Tax Collection," *Economic Journal*, Vol. 114 (July), pp. 569–91.

Aizenman, J., Y. Lee, and Y. Rhee, 2004, "International Reserves Management and Capital Mobility in a Volatile World: Policy Considerations and a Case Study of Korea," NBER Working Paper No.10534, (Cambridge, Massachusetts: National Bureau of Economic Research).

Arensberg & Harry W. Pearson (eds.), *Trade and Market in the Early Empires: Economies in History and Theory*, New York, The Free Press.

Arthur, Brian and Geoffrey McNicoll.1975. "Large Scale Simulation Models in Population and Development." *Population and Development* 1: 251-265.

Ashraf, Quamrul, and Oded Galor. "Dynamics and stagnation in the Malthusian".

Ashraf, Quamrul H., David N. Weil, and Joshua Wilde. "The effect of fertility reduction on economic growth." *Population and Development Review* 39, no. 1 (2013): 97-130.

Ayres, C. E. *et al.*, 1964. *Institutional Economics: Veblen, Commons and Mitchell Reconsidered.*

Balance-of-Payments Problems," *American Economic Review*, Vol. 89 (June), pp. 473–500.

Baldwin (ed.), *Kropotkin's Revolutionary Pamphlets*, New York, Benjamin Blom.

Baldwin, R. N. (1968). 'Note on the Editing of the Pamphlets', in Roger N. Baldwin (ed.), *Kropotkin's Revolutionary Pamphlets*, New York, Benjamin Blom.

Becker, Sascha O., Francesco Cinnirella, and Ludger Woessmann. "The trade-off between fertility and education: Evidence from before the demographic transition." *Journal of Economic Growth* 15, no. 3 (2010): 177-204.

Ben-Bassat A., and D. Gottlieb, 1992, "Optimal International Reserves and Sovereign Risk," *Journal of International Economics* , Vol. 33, pp. 345–62.

Berger, Peter L. and Tomas Luckman, 1991 (1966). *The Social Construction of Reality*. Anchor Books.

Bevir, M. (1999). *The Logic of the History of Ideas*, Cambridge, Cambridge University Press.

Birdsall, Allen C. Kelley, and Steven W. Sinding, eds., *Population Matters: Demographic Change, Economic Growth, and Poverty in the Developing World*, pp. 137-164. Oxford: Oxford University Press.

Birdsall, Nancy. Allen C. Kelley, and Steven W. Sinding, eds. 2001. *Population Matters: Demographic Change, Economic Growth, and Poverty in the Developing World*. Oxford; Oxford University Press.

Bloom, David E., and David Canning. "Global Demographic Change: Dimensions and Economic Significance." *Population and Development Review* 34 (2008): 17-51.

Bloom, David and Jeffery G. Williamson. 1998. "Demographic Transitions and Economic Miracles in Emerging Asia." *The World Bank Economic Review* 12: 419-455.

Bongaarts, John. 1978: "A Framework for Analyzing the Proximate Determinants of Fertility." *Population and Development Review* 4: 105-132.

Bongaarts, John. 1992: "Population Growth and Global Warming." *Population and Development Review* 4: 105-132.

Bookchin, M. (1972). *Post-Scarcity Anarchism* , Montreal, Black Rose Books.

Borgström Hansson, Carina, 2003. *Misplaced Concreteness and Concrete Places: Critical Analyses of divergent Discourses on Sustainability* . Lund Studies in Human Ecology 7. Lund Univeristy,

Boserup, Ester. 1981. *Population and Technological Change: A Study of Long-Term Trends*. Chicago: University of Chicago Press.

Broader version of institutional economics', *METU Studies in Development*, 23 (4).

Bryant, R., 1980. "A Model of Reserves, Bank Runs, and Deposit Insurance," *Journal of Banking and Finance*, Vol. 4, pp.335–44.

Caldwell, John. 1976. "Toward a Restatement of Demographic Transition Theory." *Population and Development Review* 2(3): 321-366.

Calvo, G., 1998. "Capital Flows and Capital-Market Crises: The Simple Economics of Sudden Stops," *Journal of Applied Economics* (November).

Calvo, G., and E. Mendoza, 2000, "Contagion, Globalization, and the Volatility of Capital Flows," *Capital Flows and the Emerging Economies* , ed. by Sebastian Edwards, (Chicago: University Chicago Press).

Catton, William. 1980. Overshoot: The Ecological basis of Revolutionary Change. Urbana: University of Illinois Press.

Century European Anarchism , London, Croom Helm. *China*, Vol.16 (October).

Cleland, John and Christopher Wilson. 1987. "Demand Theories of Fertility Transition: An Iconoclastic View." *Population Studies* 41: 5-30.

Coale, Ansley and Edgar M. Hoover. 1958. *Population Growth and Economic Development in Low Income Countries*. Princeton: Princeton University Press.

Cohen, H. ed. (1927). *Proudhon's Solution of the Social Problem* , New York, Vanguard Press.

Cohen, Joel E. 1995. *How Many People Can the Earth Support? New York: W.W. Norton.*

Cole, G.D.H. (1953). *Introduction to Economic History 1750-1950* , London, Macmillan & Co.

Cole, G.D.H. (1964). *A History of Socialist Thought*, Volume II, London, Macmillan & Co.

Cole, G.D.H. (1965). 'Socialist Thought: The Forerunners 1789-1850', in G.D.H. Cole, *History of Socialist Thought*Volume 1, London, Macmillan & Co.

Colin Hines, Andrew Kimbrell, David Korten, Sara Larrain, Helena Norberg-Hodge, Simon Retallack, Vandana Shiva, Victoria Tauli-Corpuz, Lori Wallach). 2002. Alternatives to Globalization. A Better World is Possible. Berret-Koehler, San Francisco.

Connolly, William E. 1993 (1974). *The Terms of Political Discourse*, Blackwell, Oxford.

Corcoran, P. E. (1983). 'Introduction', in Paul E. Corcoran (ed.), *Before Marx: Socialism and Communism in France 1830-48*, London, Macmillan.

Davis, Kingsley. 1967. "Population Policy: Will Current Programs Succeed?" *Science* 158: 730-

Deutsches Institut für Wirtschaftsforschung (DIW), Berlin. Sustainability economics project. (www.sustainabilityeconomics.de)

Development Agenda: Report of the Global Thematic Consultation on Population Dynamics.

Diamond, D., and P. Dybvig, 1983. "Bank Runs, Liquidity and Deposit Insurance, *Journal of Political Economy*, Vol. 91, pp. 401–19.

Docquier, Frédéric, and Hillel Rapoport. "Globalization, brain drain, and development." *Journal of Economic Literature* 50, no. 3 (2012): 681-730.

Doepke, Matthias, and Fabrizio Zilibotti. "The macroeconomics of child labour.

Dooley, M., D., Folkerts-Landau and P. Garber, 2003. "An Essay on the Revived Bretton Woods System," NBER Working Paper No. 9971, (Cambridge, Massachusetts: National Bureau of Economic Research).

Easterlin, Richard. 1975. "An Economic Framework for Fertility Analysis." *Studies in Family Planning* 6: 54-63.

Economic Literature, Vol. 22 (March), pp. 1–39.

Economies in History and Theory, New York, The Free Press.

Edison, H., 2003. "Are Foreign Exchange Reserves in Asia Too High?" in *World Economic Outlook*, (September) (Washington: International Monetary Fund).

Edwards, S., 1983. "The Demand for International Reserves and Exchange Rate Adjustments: The Case of LDCs, 1964–1972," *Economica*, Vol. 50, pp. 269–80.

Edwards, S., 2004. "Thirty Years of Current Account Imbalances, Current Account Reversals, and Sudden Stops," *Staff Papers*, International Monetary Fund, Vol. 51, Special Issue.

Edwards, S., 2005. "Capital Controls, Sudden Stops, and Current Account Reversals", NBER Working Paper No. 11170, (Cambridge, Massachusetts: National Bureau of Economic Research).

Eff, E. A. (1989). 'History of Thought as Ceremonial Genealogy: The Neglected

Epoch." *American Economic Review* 101, no. 5 (2011): 2003-2041.

Fairclough, Norman, 1995. *Critical discourse analysis: The critical study of language.* Longman, Harlow.

Ferri, E. (1909). *Socialism and Positive Science: Darwin-Spencer-Marx* , 5th edn.,

Fleming, M. (1979). *The Anarchist Way to Socialism: Elisee Reclus and Nineteenth-*

Flint, R. (1908). *Socialism*, 2nd edn., London, Sir Isaac Pitman & Sons Ltd.

Flood, R., and P. N. Marion, 2001. "Holding International Reserves in An Era of High Capital Mobility," *Brookings Trade Forum 2001*, ed. by Collins S. M. and Rodrik D. (Washington: Brookings Institution Press).

Ford, David (editor), 1990. *Understanding Business Markets. Interaction, Relationships, Networks*.Academic Press, London.

Frenkel, Jacob, 1983. "International Liquidity and Monetary Control,"in International Money and Credit: The Policy Roles, ed. by George M. von Furstenberg, (Washington; International Monetary Fund), pp. 65 109.

From Early Warnings (European Environmental Agency). Earthscan, London.

Fullbrook, Edward (editor), 2003. *The crisis in economics. The post-autistic economics movement: the first 600 days*. Routledge, London.

Fullbrook, Edward (editor), 2004. *A Guide to What's Wrong with Economics*. Anthem Press, London.

Galor, Oded, and Omer Moav. "Natural selection and the origin of economic growth." *Quarterly Journal of Economics* 117, no. 4 (2002): 1133-1191.

Galor, Oded, Omer Moav, and Dietrich Vollrath. "Inequality in landownership, the emergence of human-capital promoting institutions, and the great divergence." *Review of Economic Studies* 76, no. 1 (2009): 143-179.

Galor, Oded, and Andrew Mountford. "Trading population for productivity: Theory and evidence." *Review of Economic Studies* 75, no. 4 (2008): 1143-1179.

Galor, Oded, and David N. Weil. "Population, technology, and growth: From Malthusian stagnation to the demographic transition and beyond." *American Economic Review* 90, no. (2000): 806-828.

Galor, Oded, and David N. Weil. "The gender gap, fertility, and growth." *American Economic Review* 86, no. 3 (1996): 374-387.

Galor, Oded, and Marc Klemp. "The biocultural origins of human capital formation." Brown University Working Paper (2014).

Galor, Oded, and Omer Moav. "Das human-kapital: A theory of the demise of the class structure." *Review of Economic Studies* 73, no. 1 (2006): 85-117.

Galor, Oded. "From stagnation to growth: Unified growth theory." In Philippe Aghion and Steven Durlauf (Eds.) *Handbook of Economic Growth* 1 (2005): 171-293.

Galor, Oded. "The demographic transition: Causes and consequences." *Cliometrica* 6, no. 1 (2012): 1-28.

Galor, Oded. *Unified Growth Theory* . Princeton University Press, 2011.

García P. S., and C. G. Soto, 2004. "Large Hoarding of International Reserves: Are They Worth It?," (manuscript), Santiago: Central Bank of Chile.

Gide C. and C. Rist (1915). *A History of Economic Doctrines*, London, George G. Harrop & Co. Ltd.

Grigg, David. 1980. *Population Growth and Agrarian Change: An Historical Perspective*. Cambridge: Cambridge University Press.

Guedes Vaz (editors), 2002. *The Precautionary Principle in the 20th Century. Late Lessons*

Halperin, R. H. (1994). *Cultural Economies: Past and Present* , Austin, University of

Hardin, Garret. 1968. "The Tragedy of the Commons." *Science* 162: 1243-1248.

Harremoës, Poul, David Gee, Malcolm MacGarvin, Andy Stirling, Jane Keys, Brian Wynne, Sofia

Hazan, Moshe, and Binyamin Berdugo. "Child labour, fertility, and economic growth." *Economic Journal* 112, no. 482 (2002): 810-828.

Heilbroner, R. L. (1979). 'Modern economics as a chapter in the history of economic

Hirschman, Charles and Stewart E. Tolnay. Forthcoming 2005. "Social Demography." In Dudley L. Poston and Michael Micklin, eds., *Handbook of Population* New York: Kluwer Academic/Plenum Publishers.

Hirschman, Charles. 1981. "The Uses of Demography in Development Planning." *Economic Development and Cultural Change* 29(April 1981): 561-575.

Hirschman, Charles. 1994. "Why Fertility Changes?" *Annual Review of Sociology* 20: 203-233.

History of Political Economy , 25: 1.

History of Political Economy, 24: 1.

Hodgson, G. M. (1999). *Economics and Utopia: Why the learning economy is not the end of history*, London, Routledge.

Holling, C. S. 1994. "An Ecologist View of the Malthusian Conflict." In Kerstin Lindahl-Kiessling and Hans Landberg, eds., *Population, Economic Development, and the Environment* Oxford: Oxford University Press).

Hope for Mankind, Boston, Published by J. Warren.

Hutchison, M., and I. Noy, 2002. "How Bad Are Twins? Output Costs of Currency and Banking Crises," forthcoming, *Journal of Money, Credit and Banking*.

Hutton, W. (1996). *The State We're In* , Fully Revised Edn., London, Vintage Books.

Hyman, R. L-S. (1977). *Gustav Landauer: Philosopher of Utopia* , Indianapolis, Hackett Publishing Co.

Industriökonomisk Institut, Rapport nr. 1. Tanum-Norli, Oslo.

Influence of Herbert Spencer on Thorstein Veblen', *Journal of Economic Issues* Vol. XXIII No.3 (September).

Institute for Economic Research Working Paper 364 (2010).

International Forum on Globalization (John Cavanagh, Jerry Mander, Sara Anderson, Debi Barker, Maude Barlow, Valden Bellow, Robin Broad, Tony Clarke, Edward Goldsmith, Randall Hayes,

Johansen, Leif, 1977. *Samfunnsökonomisk lönnsomhet*. En dröfting av begrepets bakgrunn og inhold.

Johnson, D. Gale and Ronald D. Lee, eds. 1987. *Population Growth and Economic Development: Issues and Evidence*. Madison: University of Wisconsin Press.

Joshi, Shareen, and T. Paul Schultz. "Family Planning as an investment in development: Evaluation of a program's consequences in Matlab, Bangladesh." Institute for the Study of Labour (IZA) Discussion Paper No. 2639 (2007).

Kaminsky, G., and C. Reinhart, 1999. "The Twin Crises. The Causes of Banking and

Kapp K. William, 1971 (first edition 1950). *The Social Costs of Private Enterprise*. Shocken, New York.

Kapp K. William, 1976. "The Nature and Significance of Institutional Economics", *Kyklos*, Vol. 29, No. 2

Kelley, Allen C. 1988. "Economic Consequences of Population Change in the Third World." *Journal of Economic Literature* 26: 1685-1728.

Kelley, Allen C. 2001. "The Population Debate in Historical Perspective: Revisionism Revised" In Nancy Birdsall, Allen C. Kelley, and Steven W. Sinding, eds.

Population Matters: Demographic Change, Economic Growth, and Poverty in the Developing World, pp. 24-54. Oxford: Oxford University Press.

Kelley, Allen C. and Robert M. Schmidt. 2001. "Economic and Demographic Change: A Synthesis of Models, Findings, and Perspectives" In Nancy Birdsall, Allen C. Kelley, and Steven W. Sinding, eds. *Population Matters: Demographic Change, Economic Growth, and Kropotkin's Revolutionary Pamphlets*, New York, Benjamin Blom.

Kern, R. W. (1978). *Red Years Black Years: A Political History of Spanish Anarchism 1911-1937*, Philadelphia, Institute for the Study of Human Issues.

Khalil, E. L. (1995). 'Has Economics Progressed? Rectilinear, Historicist, Universalist, and Evolutionary Historiographies', *History of Political Economy*, Vol.27 No.1 (Spring).

Klemp, Marc, and Jacob Louis Weisdorf. "Fecundity, Fertility and Family Reconstitution Data: The Child Quantity-Quality Trade-Off Revisited." CEPR Discussion Papers No. 9121 (2012).

Kohler, Hans-Peter and Jere R. Behrman. " Population dynamics: Prioritizing the Post-2015

Korten, David C. 2001. *When Corporations Rule the World* (Second edition). Berret-Koehler, San

Kravis, I.B., 1984. "Comparative Studies of National Incomes and Prices," *Journal of*

Kropotkin, P. (1972a). *Mutual Aid: A Factor of Evolution*, New York, New York University Press.

Kropotkin, P. (1968a). 'Anarchist Communism: Its Basis and Principles', in Roger N.

Kropotkin, P. (1968b). 'Modern Science and Anarchism', in Roger N. Baldwin (ed.),

Kropotkin, P. (1968c). 'Anarchist Communism', in Roger N. Baldwin (ed.) *Kropotkin's Revolutionary Pamphlets*, New York, Benjamin Blom. Kropotkin, P. (1968d). 'Anarchism: Its Philosophy and Ideal', in RogerN.Baldwin (ed.), *Kropotkin's Revolutionary Pamphlets*, New York, Benjamin Blom.

Kropotkin, P. (1971). *Memoirs of a Revolutionist*, New York, Dover Publications.

Kropotkin, P. (1972b). *The Conquest of Bread*, New York, New York University Press.

Kropotkin, P. (1975). 'The State: Its Historic Role', in Martin A. Miller (ed.), *SelectedWritings on Anarchism and Revolution: P.A. Kropotkin* , Massachusetts, M.I.T. Press.

Kropotkin, P. (1988). *Act for Yourselves: Articles from Freedom 1886-1907*, London, Freedom Press.

Kropotkin, P. (1992). *Words of a Rebel* , Montreal, Black Rose Books.

Kropotkin, P. (n.d.). *Fields, Factories and Workshops* , London, Thomas Nelson & Sons (first published 1898).

Kuznets, Simon. 1967. "Population and Economic Growth." *Proceedings of the American Philosophical Society* 111(3): 170-193.

LaCapra, D. (1980). 'Rethinking Intellectual History and Reading Texts', *History and Theory*, Vol.19.

Lee, J., 2004. "Insurance Value of International Reserves," IMF Working Paper 04/175, (Washington: International Monetary Fund).

Lee, Ronald, Andrew Mason, and Tim Miller. 2001. "Saving, Wealth and Population" In Nancy

Lee, Ronald. 1983a. "From Rome to Manila; How Demography has Changed in Three Decades". *International Population Conference, Manila, 1981: Proceedings and Selected Papers*, Volume 4, pp. 505-516. Manila: International Union for the Scientific Study of Population.

Lee, Ronald. 1983b. "Economic Consequences of Population Size, Structure, and Growth." IUSSP Newsletter. 17: 43-59.

Lee, Ronald. 1986. "Malthus and Boserup: A Dynamic Synthesis." In David Coleman and Roger Schofield. Eds., *The State of Population Theory* Oxford: Basil Blackwell.

Lesthaeghe, Ronald. 1983. "A Century of Demographic Change in Western Europe: An Exploration of Underlying Dimensions." *Population and Development Review* 9: 411-436.

Lindemann, A. S. (1983). *A History of European Socialism*, New Haven, Yale University Press. *Literature*, Vol.XXI (June).

Lohman, Larry, 2006. Carbon Trading. "A critical conversation on climate change, privatisation and power". Dag Hammarskjöld Foundation, Development Dialogue No. 48 (September), Uppsala.

London, Independent Labour Party. London.

Maddison, Angus. *The World Economy: A Millennial Perspective.* Organization for Economic Cooperation and Development, 2001.

Malthus, Thomas Robert. *An Essay on the Principle of Population* J. Johnson, in St. Paul's Church-yard, 1798.

Malthus, Thomas. 1960. *On Population*. Edited and introduced by Gertrude Himmelfarb. New York: The Modern Library.

Marshall, A. (1922). *Principles of Economics: An introductory volume*, 8th edn., London, Macmillan & Co.

Marshall, A. (1925a). 'The Old Generation of Economists and the New', in A. C. Pigou (ed.), *Memorials of Alfred Marshall*, London, Macmillan.

Marshall, A. (1925b). 'The Future of the Working Classes', in A. C. Pigou (ed.), *Memorials of Alfred Marshall*, London, Macmillan.

Marx, K. (1989). *A Contribution to the Critique of Political Economy*, New York, International Publishers.

Marx, K. (1963). *Capital*, Volume 1, Moscow, Foreign Languages Publishing House.

Mason, Karen O. 1997. "Explaining Fertility Transitions." *Demography* 34: 443-54.

McBriar, A. M. (1987). *An Edwardian Mixed Doubles: the Bosanquets versus the Webbs. A Study in British Social Policy 1890-1929* , Oxford, Clarendon Press.

McCloskey, D. N. (1983). 'The Rhetoric of Economics', *Journal of Economic*

McLennan, D. (1980). *Marx before Marxism*, 2nd edn., London, Macmillan.

Mingione, E. (1991). *Fragmented Societies: A Sociology of Economic Life Beyond the Market Paradigm*, Oxford, Basil Blackwell.

Mishan Ezra J. 1980. "How Valid are Economic Evaluations of Allocative Changes?"*Journal of*

Mishan, Ezra J. 1971. *Cost-Benefit Analysis*. Allen & Unwin, London.

Moav, Omer. "Cheap Children and the Persistence of Poverty," Economic Journal 115 (2005): 88-110.

Morgan, S. Philip. 2003. 'Is Low Fertility a Twenty-First Century Demographic Crisis?" Demography 40: 589-603.

Mountford, Andrew, and Hillel Rapoport. "The brain drain and the world distribution of income." *Journal of Development Economics* 95, no. 1 (2011): 4-17.

Mountford, Andrew. "Can a brain drain be good for growth in the source economy?" *Journal of Development Economics* 53, no. 2 (1997): 287-303.

Murphy, Tommy E. "Old habits die hard (sometimes) can département heterogeneity tell us something about the French fertility decline?" Bocconi University Innocenzo Gasparini.

Murtin, Fabrice. "Long-term determinants of the demographic transition, 1870–2000." *Review of Economics and Statistics* 95, no. 2 (2013): 617-631.

Myrdal, Gunnar, 1975 (first edition 1972). *Against the Stream. Critical essays on economics*. Random House, New York.

Myrdal, Gunnar, 1978. "Institutional Economics", *Journal of Economic Issues*, Vol. 12, No.4 (December),

National Academy of Sciences. 1971. *Rapid Population Growth: Consequences and Policy Implication*. Two volumes. Baltimore.: Johns Hopkins Press.

National Academy of Sciences. 1986. *Population Growth and Economic Development: Policy Questions*. Washington, D.C.: National Academy Press.

Nettlau, M. (1996). *A Short History of Anarchism* , London, Freedom Press.

North, D. C. (1977). 'Markets and Other Allocation Systems in History: The Challenge of Karl Polanyi', *Journal of European Economic History*, Vol.6, No.3 (Winter).

Oded Stark (Eds.) *Handbook of Population and Family Economics* 1 (1997): 349-430.

Palgrave's Dictionary of Political Economy (1925). Vol.1, Henry Higgs (ed.), London, Macmillan.

Pebley, Anne R. 1998. "Demography and the Environment." *Demography*, Vol. 35: 377-389.

Pickowicz, P. (1990). 'The Chinese anarchist critique of Marxism-Leninism', *Modern*

Polanyi, K. (1957a). 'The Economy as Instituted Process', in Karl Polanyi, Conrad M.

Polanyi, K. (1957b). 'Societies and Economic Systems', in Karl Polanyi, *The Great*

Polanyi, K. (1957c). 'The Place of Economics in Societies', in Karl Polanyi, Conrad M. Arensberg & Harry W. Pearson (eds.), *Trade and Market in the Early Empires:*

Polanyi, K. (1977). *The Livelihood of Man*, New York, Academic Press.

Polanyi-Levitt, K. (1994). 'Karl Polanyi as Socialist', in Kenneth McRobbie (ed.), *Humanity, Society and Commitment: On Karl Polanyi*, Montreal, Black Rose Books.

Porter Abbot H. 2002. *The Cambridge Introduction to Narrative* . Cambridge University Press, Cambridge.

Poverty in the Developing World, pp. 67-105. Oxford: Oxford University Press. pp. 771-783.

Prasad, E., and S-J Wei, 2005, "The Chinese Approach to Capital Inflows: Patterns and Possible Explanations," IMF Working Paper 05/79 (Washington: International Monetary Fund).

Preston, Samuel H. 1987. 'The Social Sciences and the Population Problem." *Sociological Forum* 2: 619-644.

Prisman, E., M. Solvin, and M. Sushka, 1986, "A General Model of The Banking Firm Under Conditions of Monopoly, Uncertainty and Recourse," *Journal of Monetary Economics*, Vol. 17 (2): pp. 293–304.

Proudhon, P.-J. (1970). *What is Property? An Inquiry into the Principle of Right and of Government* , New York, Dover Publications.

Psacharopoulos, George. "Benefits and Costs of the Education Targets for the Post-2015 Development Agenda." Report for the Copenhagen Consensus Center (2014). Rosenzweig, Mark R., and Kenneth I. Wolpin. "Testing the quantity-quality fertility model:

Puskas Nordin, Anna-Carin, Peter Söderbaum, 2004. "Regionala aktörers tankar om hållbar utveckling. Kalmar och Uppsalaregionerna som exempel" ("houghts about Sustainable Development by regional actors. The regions of Uppsala and Kalmar as cases.") Mälardalen University, School of Business, Report 2004: 1. Västerås.

Read, D. (1994). *The Age of Urban Democracy: England 1868-1914* , New York, Longman.

regulation." *American Economic Review* 95, no. 5 (2005): 1492-1524.

Ricoeur, Paul, 1981. *Hermeneutics and the Human Sciences* . Cambridge University Press, Cambridge.

Rocker, R. (1956). *The London Years* , London, Robert Anscombe & Co.

Roll, E. (1973). *A History of Economic Thought* , 4th edn., London, Faber & Faber Ltd.

Rosenzweig, Mark R., and Junsen Zhang. "Do population control policies induce more human capital investme ? Twins, birth weight and China's "one-child" policy." *Review of Economic Studies* 76, no. 3 (2009): 1149-1174.

Russell, B. (1970). *Roads to Freedom: Socialism, Anarchism and Syndicalism*, London, Unwin Books.

Samuel, R. (1992). 'Reading the Signs II: Fact-grubbers and mind-readers', *History*

Samuelson, P.A. (1994). "Facets of Balassa-Samuelson Thirty Years Later," *Review of International Economics*, Vol. 2, No. 3, pp. 201–26.

Schabas, M. (1992). 'Breaking Away: History of Economics as History of Science',

Schultz, T. Paul. "Changing world prices, women's wages, and the fertility transition:

Schultz, T. Paul. "Demand for children in low income countries." In Mark R. Rosenzweig and

Schultz, T. Paul. "Population policies, fertility, women's human capital, and child quality." In Paul Schultz and John A. Strauss (Eds.) *Handbook of Development Economics* 4 (2008): 3249-3303.

Schumpeter, J. A. (1972). *History of Economic Analysis*, London, George Allen & Unwin.

Shiva, Vandana. 2006. *Earth Democracy. Justice, Sustainability and Peace*. Zed Books, London.

Simon, Julian. 1981. *The Ultimate Resource*. Princeton: Princeton University Press.

Skinner, Q. (1988). 'A reply to my critics', in James Tully (ed.), *Meaning and Context: Quentin Skinner and his Critics*, Cambridge, Polity Press.

Social Economy, 47: 3 (Fall).

Söderbaum Peter, 2000. *Ecological Economics. A Political Economics Approach to Environment and Development*. Earthscan, London.

Söderbaum, Peter, 2001. "Political Economic Person, Ideological Orientation and Institutional Change. On Competition between Schemes of Interpretation in Economics, *Journal of Interdisciplinary Economics*, Vol. 12, pp. 179-197.

Söderbaum, Peter, 2004a. "Decision Processes and Decision-making in Relation to Sustainable Development and Democracy – Where Do we Stand?" *Journal of Interdisciplinary Economics*, Vol. 14, pp. 41-60.

Söderbaum, Peter, 2004b. "Nobelpriset i ekonomi hinder för nytänkande" ("The Nobel Prize in Economics – Barrier for New Thinking"), *Dagens Nyheter*, Debatt, 10 oktober 2004.

Söderbaum, Peter, 2006. "Democracy and Sustainable Development – What is the Alternative to Cost-Benefit Analysis?" *Integrated Environmental Assessment and Management* , Vol. 2, No. 2, pp.182-190.

Söderbaum, Peter, 2007. "Issues of paradigm, ideology and democracy in sustainability assessment", *Ecological Economics* , Vol. 60, pp. 613-626.

Stanfield, J. R. (1986). *The Economic Thought of Karl Polanyi: Lives and Livelihood* London, Macmillan.

Stanfield, J. R. (1989). 'Karl Polanyi and Contemporary Economic Thought' *Review of*

Stark, W. (1994). *History and Historians of Political Economy*, New Jersey, Transaction Publishers.

Strassman, D. (1993). 'The Stories of Economics and the Power of the Storyteller',

Sunstein, Cass R. 2002. *Risk and Reason. Safety, Law and the Environment* . Cambridge University Press, Cambridge.

Sweden, 1860-1910." *Journal of Political Economy* 93, no. 6 (1985): 1126-1154.

Tezel, G. (1996) 'Exploring the theories of Polanyi and Chayanov: Groundwork for a the use of twins as a natural experiment." *Econometrica* 48, no. 1 (1980): 227-240.

Tierney, John.1990. "Betting the Planet." New York Times Magazine, Dec. 2, 1990, pp 52-.

Toulmin, Stephen. 1990. *Cosmopolis. The hidden agenda of modernity* . The University of Chicago Press, Chicago.

UN Development Agenda on Population Dynamics, " Report for the Copenhagen Consensus Center (2014).

UNFPA, UNDESA, UN-HABITAT, IOM (2013). Population Dynamics in the Post-2015

United Nations, 2005. *Millenium Development Goals* (www.un.org/millenniumgoals)

United Nations. 1973 (1953). *The Determinants and Consequences of Population Trends New Summary of Findings on Interaction of Demographic, Economic and Social Factors*. Population Studies, No. 50. New York: United Nations.

Vandervelde, E. (1907). *Collectivism and Industrial Evolution*, London, Independent

Warren, J. (1863). *True Civilization: an Immediate Necessity and the Last Ground of*

Watkins, Susan C. 1991. *From Provinces into Nations: Demographic Integration in Western Europe, 1870-1960.* Princeton: Princeton University Press:

Welford, Richard, (editor), 1996. *Corporate Environmental Management. Systems and Strategies* Earthscan, London.

Williams, R. (1988). *Keywords: A vocabulary of culture and society*, London, Fontana

Woodcock, G. & I. Avakumovic (1970). *The Anarchist Prince: A Biographical Study of Peter Kropotkin*, New York, Kraus Reprint Co. *Workshop Journal*, No.33.

World Commission on Environment and Development (WCED), 1987. *Our Common Future*. Oxford University Press, Oxford.

Wrigley, E.A. 1969. *Population and History*. New York: McGraw Hill.

York, Richard, Eugene Rosa, and Thomas Dietz. 2003. "Footprints in the Earth: The Environmental Consequences of Modernity." *American Sociological Review* 68: 279-300.

Index

www.ingramcontent.com/pod-product-compliance
Ingram Content Group UK Ltd.
Pitfield, Milton Keynes, MK11 3LW, UK
UKHW020245290926
15470UKWH00002B/745

9 789387 057579